CW00820416

Into the Belly of the Beast

Exploring London's Victorian Sewers

Into the Belly of the Beast
Exploring London's Victorian Sewers

Paul Dobraszczyk

Spire Books Ltd

PO Box 2336. Reading RG4 5WJ
www.spirebooks.com

Spire Books Ltd
PO Box 2336
Reading RG4 5WJ
www.spirebooks.com

CIP data:
A catalogue record for this book is available
from the British Library
ISBN 978-1-904965-24-4

Designed and produced by John Elliott
Text set in Bembo

Publication of this book was made possible by generous grants from the Society of Architectural Historians of Great Britain, the University of Reading, the Marc Fitch Fund, and the Paul Mellon Centre for Studies in British Art.

Contents

Preface

Walking along the elevated Greenway – in effect, the top of the northern outfall sewer – between High street in Stratford and Manor road in West Ham, presents one of the bleakest scenes in east London – an area scarred by the ravages of industry and piecemeal development. Yet, the sight of the Abbey Mills pumping station provides a startling contrast. With its fantastical central lantern, turreted towers and multicoloured brickwork, it resembles a cathedral in the midst of an otherwise architecturally barren landscape. I first came across this building whilst working as an administrator for Thames Water plc, after completing my undergraduate degree in 1998. Even then I found it an enigmatic building. What could its flamboyant style have to do with its function, that is the pumping of human excrement? Other employees with engineering expertise told me the lavish ornament was either superfluous or organised according to the practical function of the building. I found this explanation unsatisfactory and embarked upon the doctoral research in 2003 that would eventually result in this book.

The questions generated by Abbey Mills led to a wider consideration of the spaces of London's sewerage system, to which the former are intimately connected. In doing so, I uncovered a veritable mass of images that had previously been either ignored by historians or simply used as illustrations in other books about the project. If the vast spaces of the sewers beneath London are now invisible to its present-day citizens, this was not the case during its planning and construction in the mid-Victorian period. Then, it was visualised almost obsessively: in maps, plans, photographs, newspaper illustrations, and in the architecture of its pumping stations. As will become clear in this book, it was this field of imagery that became the central focus of my research, around which the initial questions about Abbey Mills hinged.

I could not have completed the research without the assistance of many scholars, fellow students and friends. Throughout the period of research I was supported with help and advice, particularly that from Chris Pierce, who was generous in both his enthusiasm and his time for long lunches in Oxford and London. Alongside him, my supervisor Paul Davies in the Department of History of Art and Architecture at the University of Reading provided ample encouragement and close attention to the progress of my research, particularly its architectural elements. Likewise, John Elliott offered advice and support when especially needed and, with Geoff Brandwood, provided me with the opportunity to publish the research with Spire Books, made possible by generous grants from the Marc Fitch Fund and the Paul Mellon Centre. I am grateful to the many other academics who read and commented on various portions of the research, particularly my examiners Mark Jenner and Sue Malvern, but also Judi Loach and Barrie Bullen for comments on chapter 5 and Gloria Clifton and the anonymous reviews for *Technology and Culture* who provided detailed comments on an earlier version of chapter 3. Other invaluable help along the way has been provided by Christine Bainbridge, Philip Barker, Jeff Brookes, Inga Bryden, Bryce Caller, Ben Campkin, Rosie Cox, Jo Mordaunt Crook, Richard Dennis, Roddy Dewe, Dale Dishon, Lisa Dobraszczyk, Tadzio Dobraszczyk, Nick Driver, Paul Driver, Michael Dunmow, Anne Farrington, Stephen Halliday, Anna Hines, Quintin Lake, Carol Morgan, Margaret Morgan, Liza Pickard, Lucy Porten, Tim Proctor, Malcolm Robinson, Deborah Sanders, Renato Sansa, Mike Seabourne, Niall Scott, Rachel Stewart, Sharyn Sullivan-Tailyour, Robert Thomas, Volker Welter, and Matthew Wood. Special thanks to the staff at Reading University Library and the Bodleian Library (upper reading and map rooms especially), the British Newspaper Library, the Guildhall Library, and the London Metropolitan Archives for their assistance with the illustrative and primary source material. Grateful thanks also to students who braved two new courses I taught at the Department of Continuing Education at the University of Reading: 'The Slum and the Sewer: Sanitation and Space in Mid-Victorian London' in Summer 2004; and 'Technology and Art: Victorian Industrial Architecture' in Spring 2006. They restored my sanity and helped me communicate my research to a wider audience.

I was able to begin the research thanks to substantial funding in the form of a Jonathan Vickers Bursary from the Society of Architectural Historians of Great Britain and a Reading University Faculty Studentship. The shape of the work gained enormously from my presentation of work in progress at postgraduate seminars at the Centre for Urban History, University of Leicester (2003) and the University of Reading (2003 and 2005), the Literary London conference (2004), the fifth annual conference of the British Association of Victorian Studies (2004), the seventh and eighth International Conferences on Urban History (2004 and 2006), the third global conference, 'Monsters and the Monstrous: Myths of Enduring Evil' (2005), the 'Architecture and Dirt' session at the 60th Annual Meeting of the Society of Architectural Historians (2007), and the Centre for Metropolitan Studies at the University of London (2007). An earlier version of chapter 1 has been published as 'Mapping Sewer Space in Mid-Victorian London', in Ben Campkin and Rosie Cox (editors), *Dirt: New Geographies of Dirt and Purity* (IB Tauris, London, 2007); chapter 3 as 'Image and Audience: Contractual Representation and London's Main Drainage System', *Technology and Culture* 49 (2008); chapter 4 as 'Sewers, Wood Engraving and the Sublime: Picturing London's Main Drainage System in the *Illustrated London News*, 1859-62', *Victorian Periodicals Review* 38: 4 (2006); chapter 5 as 'Historicizing Iron: Charles Driver and the Abbey Mills Pumping Station (1865-68)', *Architectural History* 49 (2006); chapter 6 as '"Monster Sewers": Experiencing London's Main Drainage System', in Niall Scott (editor), *Monsters and the Monstrous: Myths of Enduring Evil* (Rodopi, Amsterdam and New York,

2007); and chapters 5 and 6 as 'Architecture, Ornament and Excrement: the Crossness and Abbey Mills Pumping Stations', *Journal of Architecture* 12: 4 (2007).

Finally, my greatest debt is to Robin Winters, who I first met at Thames Water in 1998, and who has been a constant source of encouragement ever since. Without his generous provision of source material, supportive and corrective advice, and unrestricted access to Abbey Mills, this book would never have seen the light of day. Now retired from many years of dedicated service to both Thames Water and its archives, it is to him that I dedicate this volume.

Paul Dobraszczyk
Oxford
June 2009

Picture Credits

Illustrations are reproduced by permission of:

Guildhall Library, Corporation of London: 2.1, 2.2, 4.1(b), 4.7

City of London, London Metropolitan Archives: 1.1 (RM 28/9); 1.3 (SC/OS/LN/20/031); 1.4 (MCS/489/45); 1.5 (MCS/P/25/1-15 (12)); Pl. II (LCC/CE/MD/09/005); Pl. III (MCS/478/5); Pl. IV (a,b) (MCS/479/13); Pl. V (MBW/2508)

The British Library, London: 5.13

The Bodleian Library, Oxford: 5.9(d), Pl. VIII(c), Pl. VIII(d), 5.14(b), 5.16(c), 5.17(c)

Thames Water plc: I.1, I.2, Pl. I, 1.2, 2.3, 3.1, 3.2, 3.3, 3.5, 5.2, 5.4, 5.7, 5.8, Pl. VI(a), 5.9(a), 5.9(b), 5.9(c), Pl.VII(a), 5.10(a), Pl.VIII(a), 5.12(c), 5.12(d), 5.14(a), 5.15(a), 5.15(b), 5.16(a), 5.16(b), 5.18(a), 5.18(b), 5.18(c), P.1

Manchester City Art Galleries: 4.5

Getty Images: 4.3(b)

Quinin Lake: jacket front photograph

Introduction

Monstrous or marvellous? The creation of London's Victorian sewerage system – known as the main drainage system – was, like the building of the railways, a wonder of its age. It was on a scale almost unimaginable a century before and carried with it a mighty public purpose: to sanitise what was then the greatest city on earth. The work was highly visible and, not surprisingly, attracted enormous interest from London's press, who gave ecstatic praise to this 'the most extensive and wonderful work of modern times'.[1]

The year 1861 also saw the publication of several articles about the London sewers in Charles Dickens's journal *All The Year Round,* written by the journalist John Hollingshead (1827-1904).[2] The article published on 17 August discusses the main drainage system, then under construction in the city. The language of Hollingshead's article generally mirrors the celebratory tone of the press accounts and all seems well until the last page where some startling observations are made: according to Hollingshead, if some regarded the main drainage system as a 'great plan', others saw the new sewers as 'volcanoes of filth; gorged veins of putridity; ready to explode at any moment in a whirlwind of foul gas, and poison all those whom they fail to smother'.[3]

Why did some think London's new sewers wonderful, others monstrous? Are there yet more ways in which they were perceived? How were these different viewpoints expressed and how did they relate to one another, if at all? What might the answers to these questions say about the wider cultural impact of the main drainage system? The construction of London's new sewers may be seen by many as culturally-neutral achievement that was implemented in an atmosphere of unambiguous public appreciation, but, as Hollingshead hints at, there is another story to be told, one that gets behind his startling comments and investigates other perceptions of its spaces. This book will tell that story.

The Main Drainage System

Before discussing the existing literature on the main drainage system in more detail, I want briefly to define the object of this study – that is, what it is and what it is not. The period covered by this book – 1848 to 1868 – represents the years when London's new system of sewers was first planned (1848-58) and then constructed in its first and most important phase (1859-68). Designed by the engineer Sir Joseph Bazalgette (1819-91), the main drainage system consists of five distinct large-scale sewers that cross London either side of the Thames from west to east, converging in outfall sewers running roughly parallel to the river (**I.1**). These are known as 'intercepting' sewers in that they intercept waste from existing street sewers and divert it to outfalls on the Thames outside the city limits. The sewage was originally discharged into the river at high tide to prevent it from flowing back into the city area. Included in the main drainage system are four principal pumping stations, situated at points in the system where the sewage needed to be lifted from low-lying areas up to the level of the river.

The main drainage system, thus defined, must be differentiated from other terms that I will use throughout this book: first, 'drainage' refers to any method of collecting waste from individual properties, groups of properties, or areas of land, usually by means of underground channels known as 'drains'; second, the term 'sewer' refers to any channel that collects waste from the much smaller drains; third, 'sewerage' describes any system of sewers used to drain entire communities, while 'sewage' refers to the contents of the sewers themselves; finally, 'sanitation' encompasses all of these terms and is defined as any subject dealing with sanitary matters, principally the collection and disposal of waste matter. The main drainage system, then, is a larger intervention than a drain or a sewer but a smaller component of sewerage or sanitation. To clarify these distinctions, it might be useful, like Victor Hugo in his celebrated description of the sewers of Paris in *Les Misérables* (1862), to imagine London's underground infrastructure as resembling the form of a tree: the smallest twigs representing the household drains; the larger branches the street sewers; the largest branches and trunk the main drainage system; with the whole arrangement of twigs, branches and trunk representing the city's complete sewerage system.[4]

Despite making some references to the wider sanitary infrastructure of London and other cities, the primary focus of the book is the main drainage system itself. Consequently, there are many related topics that fall outside the scope of the book. These include: first, domestic or 'private' sanitation – that is, individual drains or sanitary appliances within the house;[5] second, other 'public' or main sewers that make up the bulk of London's sewerage system; and third, associated projects with the main drainage system, such as the Thames Embankment which, although containing part of the main drainage system, was a large-scale civil engineering project in its own right.[6] Other related subjects include the fate of the London sewage itself and debates about both its content and disposal, which formed an important background to discussions about the main drainage system from the 1840s onwards (see chapter 2) and, last, the workers associated with the main drainage system, such as the sewermen and maintenance staff at the pumping stations.[7] The latter topic, in particular, remains a fruitful area for new research that might follow the example of Donald Reid's important study of the Paris sewers and their associated workers, *Paris Sewers and Sewermen: Representations and Realities* (1991).[8]

Histories of London's Sewers

The built environment of our towns and cities is made up of a vast array of different structures, from architectural monuments to the most mundane of buildings. This environment usually extends below ground as what we more commonly call 'infrastructure', made up of pipes, cables, tunnels and sewers. All the components of this built environment, whether visible or invisible to the public, have both a 'maker' – or 'makers' – and also an 'audience', that is, those who are affected by what is built. In short, all spaces are both produced and received, whether consciously or otherwise. The construction and reception of the most mundane structures of the city, such as sewers, is dictated not only by physical, economic and political factors but also by beliefs, values and ideas. Even if, today, sewers are a subject that mainly occupy the minds of specialist engineers, in the mid-19th century they were almost a national obsession and were related to a host of concerns, some of them technical but others that were moral, utopian or even theological. And yet, within conventional historical accounts of the main drainage system the relationships between its technical aspects and the wider ideas that informed these have been largely overlooked.

Within the large body of literature that deals with London's sanitary development in the 19th century, we would expect the main drainage system to figure prominently. Henry Jephson's *The Sanitary Evolution of London* (1907) remains the most important early source in this regard: Jephson describes the development of the main drainage system in the context of a wide-ranging history of London's sanitation.[9] Subsequent accounts have followed a similar approach, placing the development of London's sewers within a variety of historical contexts, such as public health, the city's government, the river Thames, or the often-convoluted debates amongst sanitary reformers and engineers in the 1840s and 1850s.[10] Although providing a rich variety of contexts for the main drainage system, most of these accounts make only cursory references to its possible wider cultural impact. With a more sustained focus on the main drainage system itself, Stephen Halliday's book, *The Great Stink of London: Sir Joseph Bazalgette and the Cleansing of the Victorian Metropolis* (1999), has the overriding aim of memorialising Bazalgette, that is, to celebrate his 'heroic' achievements. Consequently, although it includes a rich range of source material as illustrations, it does not investigate this material in itself or question and enlarge upon how the main drainage system was represented, a deficiency that I readily redress throughout this book.

London's sewers also take centre stage in more specialist biographies of the city – ones that deal with its subterranean structures. Beginning with Hollingshead's *Underground London*, published in 1862, subterranean London has provided a consistently fascinating site for historians interested in the 'hidden' life of the city.[11] The most celebrated of these accounts is Richard Trench and Ellis Hillman's *London Under London: a Subterranean Guide*, reprinted eleven times since 1984 and revised in 1993. An encyclopaedic guide to London's hidden 'sights' – including gas and water pipes, tunnels under the Thames, abandoned defences, and lost Underground stations – Trench and Hillman's study situates these within a broader history of the city's underground world. The main drainage system features prominently, as a site of both technological efficiency and hidden dangers. The authors' journey into one of Bazalgette's sewers – described in the chapter titled 'The Bowels of the Earth' – recalls those made by Hollingshead in 1862 and the observations they make raise, but

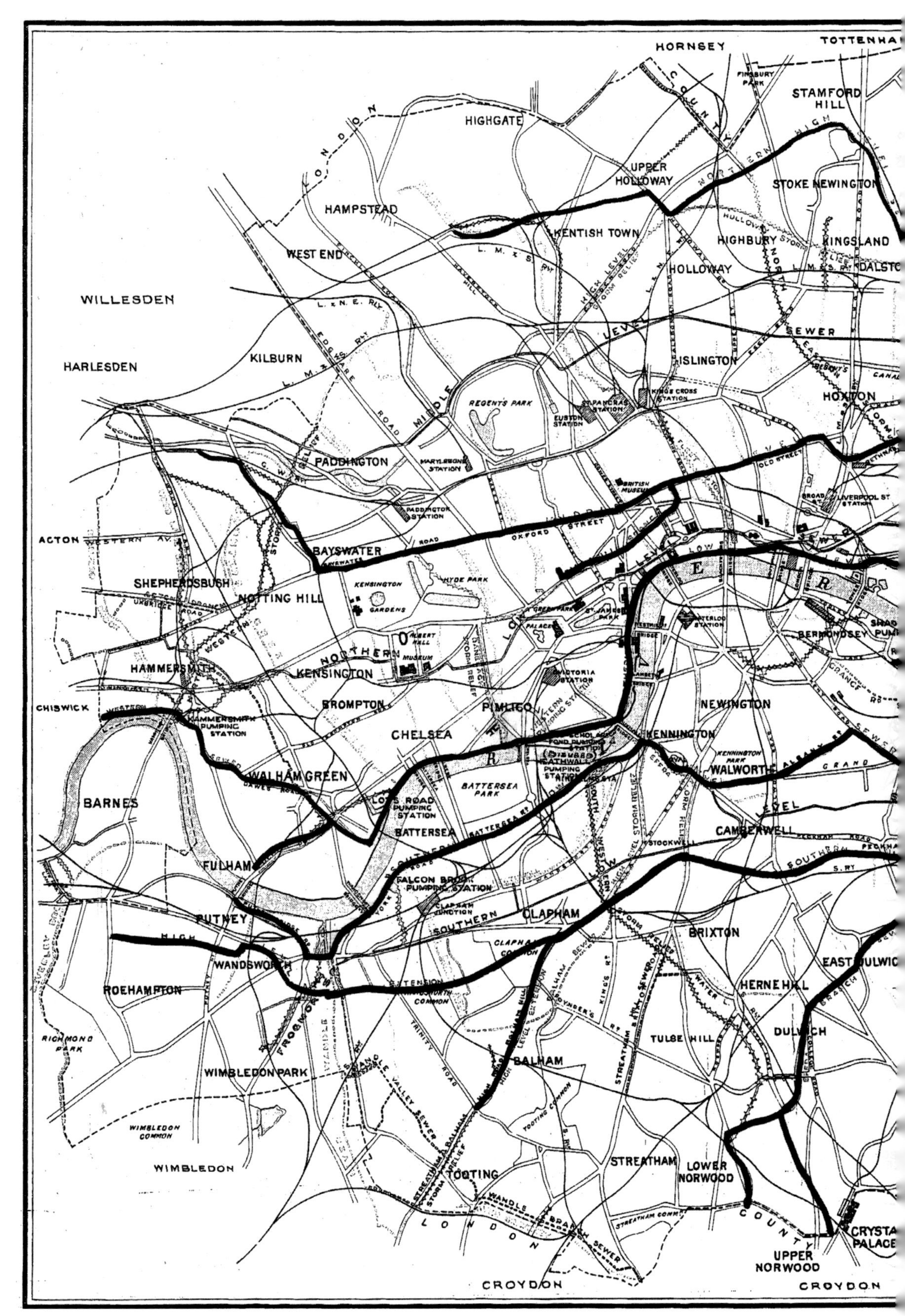

I.1 London's sewerage system in 1956, then under the control of London County Council. Bazalgette's intercepting sewers, which formed the main drainage system, are shown as the darker lines running from west to east across the city.

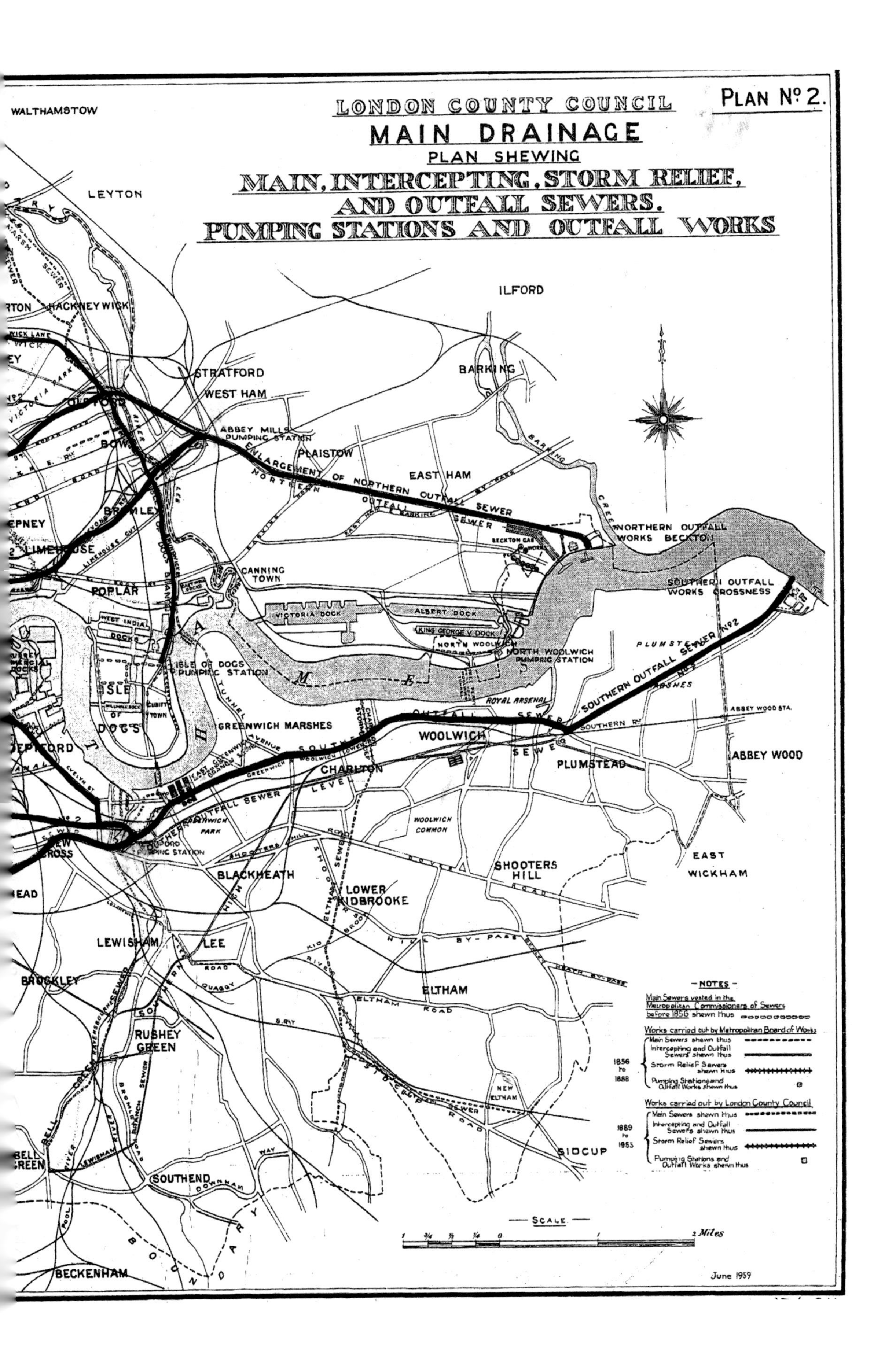

LONDON COUNTY COUNCIL
PLAN No 2.
MAIN DRAINAGE
PLAN SHEWING
MAIN, INTERCEPTING, STORM RELIEF, AND OUTFALL SEWERS.
PUMPING STATIONS AND OUTFALL WORKS
WALTHAMSTOW
LEYTON
ILFORD
HACKNEY WICK
STRATFORD
WEST HAM
BARKING
ABBEY MILLS PUMPING STATION
PLAISTOW
EAST HAM
ENLARGEMENT OF NORTHERN OUTFALL SEWER
NORTHERN OUTFALL WORKS BECKTON
SOUTHERN OUTFALL WORKS CROSSNESS
BOW
BROMLEY
POPLAR
CANNING TOWN
VICTORIA DOCK
ALBERT DOCK
KING GEORGE V DOCK
NORTH WOOLWICH PUMPING STATION
ISLE OF DOGS PUMPING STATION
ISLE OF DOGS
GREENWICH MARSHES
WOOLWICH
CHARLTON
PLUMSTEAD
ABBEY WOOD
SOUTHERN OUTFALL SEWER No 2
WOOLWICH COMMON
SHOOTERS HILL
EAST WICKHAM
BLACKHEATH
LOWER KIDBROOKE
LEWISHAM
LEE
BROCKLEY
ELTHAM
RUSHEY GREEN
NEW ELTHAM
SIDCUP
SOUTHEND
BECKENHAM
- NOTES -
Main Sewers vested in the Metropolitan Commissioners of Sewers before 1856 shewn thus
Works carried out by Metropolitan Board of Works
1856 to 1888
Main Sewers shewn thus
Intercepting and Outfall Sewers shewn thus
Storm Relief Sewers shewn thus
Pumping Stations and Outfall Works shewn thus
Works carried out by London County Council
1889 to 1955
Main Sewers shewn thus
Intercepting and Outfall Sewers shewn thus
Storm Relief Sewers shewn thus
Pumping Stations and Outfall Works shewn thus
SCALE
2 Miles
June 1959

do not answer, questions that I address in chapter 6: how and why is the main drainage system simultaneously perceived as a site of order and control and fearful disgust? What do the answers to these questions suggest about its wider cultural impact?

However, there are some more recent sources that offer the possibility of a more fruitful reading of the London's sewers in relation to their wider social and cultural contexts. David Pike's extraordinary book *Subterranean Cities: the World Beneath Paris and London, 1800-1945* (2005) explores the many meanings ascribed to sewers in London in the 19th century: as symbols of hell, poverty or crime; as a threatening space associated by the middle classes with social outcasts; a 'domestic' space sheltering those seeking refuge from the world above; a technological space defined by engineers; or a space directly associated with the figure of the prostitute.[12] Pike's approach, in bringing together such contradictory representations of these spaces, suggests a new way of approaching the main drainage system, one that I readily adopt in this book. By concentrating on *representations* of the main drainage system – whether pictorial or written – spanning the period when it was planned and constructed, this book will bring out the different, and often contradictory, ways in which its spaces were understood. Complementing Pike's primary focus on literature, Lynda Nead's book *Victorian Babylon – People, Streets and Images of Nineteenth-century London* (2000), focuses on visual representations – from the work of established artists to wood engravings in the illustrated press. Nead argues that images depicting the construction of the main drainage system in the 1860s contributed an important element in a developing 'visual culture of drains'.[13] Although providing only a brief discussion of the main drainage system, Nead makes some important observations. Like Pike, she suggests that the project – in terms of how it was represented – can be understood in many different ways: from an ordered rational intervention seen in maps, charts and engineering drawings, to a disruptive spatial intervention, seen in engravings of the construction process in the *Illustrated London News*.[14] This book, especially in chapters 1, 4 and 6, draws directly on Nead's approach, in effect enlarging on her brief but important observations and expanding her field of enquiry to a larger range of representations directly connected to the main drainage project.

Finally, Michelle Allen's recent book *Cleansing the City: Sanitary Geographies in Victorian London* (2008) goes against the grain of the existing literature on the subject by highlighting 'some of the difficulties, discomfort, and fears associated not simply with pollution but also with purification – a process we are inclined to see as generally positive'.[15] With a focus on a wide range of primary source material – from anonymous newspaper and magazine articles, pamphlets by sanitarians and engineers, to the novels of Charles Dickens and George Gissing – Allen charts an important strain of resistance to sanitary reform from the 1830s to the end of the century, part of which focuses on London's sewers. By recovering these sometimes oppositional, sometimes ambivalent responses, Allen provides insight into the contested nature of sanitary reform and also a model for my own more focused investigation into the many ways in which the main drainage system was understood and represented.

Exploring Sewer Spaces

As I have emphasised, the principal concern of this book is to investigate how London's main drainage system was represented during the period when it was planned and constructed. Such representations include not only the documents produced by the planners and engineer, but also by those who responded to it, specifically, within this book, London's press. Once completed the main drainage system remained – and still remains – largely hidden from public view because it is made up of an invisible – because underground – series of spaces (the pumping stations being the notable exception). However, during the period when it was planned and constructed these underground spaces were visualised in a variety of different media, including maps, engineers' plans and reports, contract drawings, photographs, pumping stations, newspaper articles and illustrations, and book-length studies. Indeed, this eventual fate of invisibility demanded such an initial emphasis on the visual: it enabled what would become an invisible spatial intervention to be conceived, explained, constructed and presented to the public. The pumping stations – the most visible part of the main drainage system – were both an integral part of the system and also important sites, then and now, for the promotion of the system to the public: places where the vast but invisible sewerage system is 'summed up' in a celebratory architectural statement. In chapters 5 and 6 I focus on the architectural features of the pumping stations – a corollary to the previous chapters, which discuss how the main drainage system's invisible spaces were visualised. The pumping stations served, and still serve, as important links between the invisible sewerage system and the visible city, not only in the literal sense of them as functional parts of that system but also as symbolic connections between its creators and the public.

If the pumping stations can be considered as architecture, defined conventionally as a building with artistic merit, then how might the entire main drainage system – the vast citywide network of sewers to which the pumping stations connect – be explored from an architectural perspective? Throughout the period of research, I found the notion of 'space' a fruitful way of analysing the main drainage system as whole, particularly its theoretical development in Henri Lefebvre's *The Production of Space* (1991) and Michel de Certeau's *The Practice of Everyday Life* (1984). Although these texts are often complex and philosophically dense, both theorists make the straightforward point that spaces are both conceived and perceived – that is, on the one hand they are designed by architects, planners and engineers and, on the other, they are lived in on a day-to-day basis by people and experienced in ways quite different from what the designers might have intended.[16] For both theorists, spatial production, whether at the level of a skyscraper or a sewer, should be understood within this framework. Such distinctions may seem rather obvious in regard to buildings that are inhabited in one form or another: it is easy to understand the differences between the architect and client's (or user's) viewpoint in the design and construction of a house. Yet, in relation to more everyday spaces they tend to be overlooked, especially when, in the case of sewers, those spaces are hidden from view and not 'used' according to any conventional understanding of the word. If we can easily picture a 'user' in relation to everyday spaces, such as houses or streets, how can we envisage one for the invisible space of a sewer? Given its subterranean status, does such a user exist at all? We might think of the sewer workers – those who monitor the flows and blockages within the sewers – as the most obvious 'users' of these spaces; but equally we are all users – or at least potential users – of sewers. The intricate webs of underground connecting pipes that link the individual citizen with the public sewer are, today, ubiquitous and largely taken-for-granted. We are all at once physically removed from the invisible space of the sewer but also simultaneously universally 'connected' to it in the ceaseless but unseen flows between the pipes that drain our refuse from house to street and from street to the sewage treatment works. However, in the time period under consideration in this book – 1848 to 1868 – the user of the main drainage system in this sense was only implied. Throughout the period of planning and construction, the main drainage system was not 'operational' in the way that it is today. During construction, and especially in the ceremonies marking its completion, a very different type of user experienced the project: those who witnessed and represented these events, namely London's press. My definition of 'user' in this sense is explored in more detail in chapters 4 and 6, both of which consider press responses to the construction of the main drainage system (chapter 4) and to the ceremonies marking its completion (chapter 6).

The Sanitary Development of London

The planning and construction of the main drainage system was but one event, albeit perhaps the most dramatic, in the history of London's sanitary development. Even before the rapid deterioration of London's sanitary state in the first half of the 19th century, the city's watercourses were never particularly salubrious, as testified by Ben Johnson's mock epic-poem describing a nauseating journey along the river Fleet in the early-17th century.[17] But with the population of London increasing almost threefold from 1800 to 1850, the city's sanitation, or the lack of it, became a dominating concern.[18] This exponential growth of the city's population, largely concentrated in already densely crowded areas of the city, led to serious strain being placed on a once effective and sustainable system of natural drainage. London's many rivers – tributaries of the Thames such as the Fleet, Westbourne, and Tyburn – had, up until the beginning of the 19th century, provided a ready means of draining rainwater within the built-up area.[19] The gradual expansion of this built-up area led to these rivers being systematically covered over, putting an ever-increasing strain on this existing system, which had been in place for centuries. The usual method of disposing household sewage – even up until the mid-19th century – was to empty it into pits, known as cesspools, located close to dwellings, with most households having access to one. Workers, known as 'nightmen', removed the sewage from cesspools at night and were able to dispose of it at a profit to farmers, whose fields were close by to the city limits.[20] The city's population also benefited from developments in the supply of piped water in the early-19th century. The substitution of cast iron for wood in the manufacture of water mains meant that the new London water companies were able to deliver a more regular supply at higher pressure. Coupled with the invention and subsequent popularity of the water closet, the increased availability of water resulted in much greater volumes of water and sewage being discharged into both London's rivers and its existing sewers.[21] Not surprisingly, by the 1840s, densely populated areas of London became the focus for sustained investigation by would-be sanitary reformers, particularly insalubrious areas of the city, such as the Kensington Potteries and parts of the parish of St Giles's in Westminster. These reformers, such as Edwin Chadwick (1800-90), began to investigate the detrimental effects of poor sanitation and to use the evidence they collected – often in the form of lurid descriptions – to argue the case for reform. Chadwick's interest in sanitation arose partly as a result of his work as secretary to the Poor Law Commission, established in 1834 under the Poor Law Amendment Act.[22] The new Commission attempted to reform London's administrative boundaries by setting up a centralised, government-funded body to administer poor relief to its more unfortunate citizens. The Commission overrode London's existing system of governing welfare, which was legally and administratively a loose conglomerate of some 300 individual parishes and wards.[23] From his involvement in the Commission until the publication of his *Report on the Sanitary Conditions of the Labouring Population of Great Britain* in 1842, Chadwick used his influence to focus the attention of the Commission on the relationship between poor sanitation and disease, arguing that those who required poor relief were often victims of the insanitary environments in which they lived.

The year 1848 – the starting date for this book – represents a key moment in the sanitary evolution of London and also of Chadwick's career: a new centralised governing body, the Metropolitan Commission of Sewers, headed by Chadwick, was set up in late 1847 to replace the heterogeneous group of governing bodies that had existed for centuries – the London Sewer Commissions. The Metropolitan Commission of Sewers was formed with an explicit goal: to plan and construct a new citywide sewerage system for London. Even though such a system would not be built for another decade, the year 1848 marked the genesis of the main drainage system in that it saw the emergence of a new way of seeing the city's sanitation, which I discuss at length in chapter 1. What followed in the 1850s was a protracted and sometimes vitriolic debate amongst engineers and reformers as to exactly what type of sewerage system London should have: chapter 2 focuses on the important currents within these debates and investigates the importance of representations in the presentation of the differing viewpoints and the eventual dominance of Bazalgette's conception. Once Bazalgette's main drainage system was given Parliamentary backing in 1858, it moved at an astonishing speed from concept to reality, with nearly 82 miles of intercepting sewers and three pumping stations built in the following nine years. Bazalgette's perspective on the construction process is explored in chapter 3 while public – in this case press – perceptions of the same are explored in chapter 4. Chapter 5 concentrates in some detail on the architecture of the pumping stations, focusing on the Abbey

I.2 Descending into the northern outfall sewer during Thames Water's 'Open Sewers Week', May 2007.

Mills pumping station (1865-68), built in the latter stages of the construction process but nevertheless a crucial part of it. The book concludes (chapter 6) with the completion of the system in 1865 south of the Thames and 1868 north of the river and the attendant opening ceremonies. These ceremonies represent an important threshold between construction and operation: they were events that both celebrated the permanence of the main drainage system in London's urban fabric and also marked its disappearance from public view.

The period after 1868, although outside the scope of the book, remains a little-studied period in the history of London's sanitary infrastructure.[24] In the years following the ceremony held at the Abbey Mills pumping station in 1868, London's main sewers and individual household drains were gradually connected to the main drainage system, which, until 1912, continued to be greatly enlarged.[25] These developments were accompanied by protracted debates about how best to dispose of London's sewage.[26] From the 1880s onwards, the development of increasingly sophisticated methods of treating sewage at the outfalls at Crossness and Barking meant that the question of sewage disposal lost its previous sense of urgency.[27] However, the dumping of sludge (the part of sewage that remains after treatment) in the North Sea continued until 1998, when European directives came into force that prohibited such methods of disposal. Incineration of the London sludge is now carried out at Barking and Crossness, with experiments continuing on how best to dispose of the highly toxic residual ash.[28] Even today, Bazalgette's main drainage system, although greatly enlarged and badly in need of renovation, remains a primary focus for debates on the future of London's sanitary infrastructure. The proposed Thames Tideway Tunnel, a giant 7.2-metre diameter sewer to run deep beneath the bed of the river Thames from Hammersmith to Beckton, is being touted as the 'second phase' of Bazalgette's interceptor system. Designed to prevent the regular overflow of Bazalgette's system into the Thames, it remains to be seen whether it can be completed by the proposed date of 2015 and within its estimated budget of two billion pounds.

Today, the Abbey Mills pumping station, despite being replaced by a more efficient modern counterpart, continues to have an important function as the 'public face' of the main drainage system. Thames Water plc, who now own the site and maintain the main drainage system, use Abbey Mills as the base for their annual 'Open Sewers Week', which includes a tour of the building, a descent into the northern outfall sewer – the largest of Bazalgette's intercepting sewers – and an historical lecture celebrating the main drainage system and the work of its inheritors (**I.2**). The symbolic potency of Abbey Mills, explored in detail in chapters 5 and 6, may yet continue to define public attitudes towards the hidden 'belly' of London for a long time to come.

Section I: Planning

1
Maps and Sewers

The primary way of visualising sewers is through maps where, along with other types of subterranean infrastructure, they are made visible as lines, usually coloured red, superimposed onto the spaces they lie beneath, such as roads, paths and buildings. So commonplace are such images that we invariably take for granted their apparently 'neutral' status: they serve a purely practical purpose and seem to provide us with an objective mirror of 'nature'. Yet, recent cartographical analysis – particularly the later work of John Brian Harley – has shown that maps are far from neutral images: they embody particular ways of viewing the world.[1] For Harley and others, all maps select and present information according to their own systems of priority and as a consequence present only part of the true picture of any environment.

This chapter focuses on three types of map, produced in London from 1848 to 1851, which were closely related to the future improvement of the city's sewers: first, the Ordnance Survey of London, which mapped London's above-ground topography; second, maps produced as a result of a concurrent subterranean survey of existing sewers; and third, a 'hybrid' map that attempted to combine the results of both surveys. These maps laid the conceptual and practical foundations for the main drainage system, eventually constructed in the 1860s, and are therefore of immense importance when considering the origins of the project.[2] Yet, as I show in this chapter, imagining London with a completely new system of sewers was fraught with contradictions, ones that are revealed in the maps themselves.

The Ordnance Survey

In mid-19th century London, maps played an important role in the visual culture of the city. The development of lithography from the 1820s onwards enabled them to be produced and marketed more cheaply than ever before; by 1851, a proliferation of maps, with a wide variety of different functions, would have formed a vital and accessible form of urban representation.[3] This period also saw a rise in the importance of scientific rigour in relation to mapping, which 'strove for accuracy and clarity in depicting the urban landscape'.[4] Such a change was, in part, propelled by the emergence of the Ordnance Survey, whose origins lay in the world of military strategy.[5] In the decades up to 1850, the Ordnance Survey had risen to a position of cartographic dominance over civil surveyors with the production of series of maps of Great Britain. Coupled with their emphasis on military planning, the Ordnance Survey employed new surveying techniques, culminating in the first Trigonometric Survey, begun in 1791. With the promotion of a new scientific basis for mapmaking, the work of the Ordnance Survey became a model for those who sought to imbue other disciplines with a similar method.

Perhaps the most important of these – in relation to sanitary reform – was Edwin Chadwick, the most obsessive cataloguer of dirt and disease in the mid-19th century. Chadwick's *Report on the Sanitary Conditions of the Labouring Population of Great Britain* (1842) charted in immense detail the relationship between poor sanitation and incidence of disease, employing a wide range of statistical data and eyewitness-based evidence that claimed to have a scientific origin. Sanitary cartography, developed in the later 1840s, formed not only a perfect visual complement to Chadwick's verbal reports, but also a means of depicting proposed sanitary improvements. Chadwick quickly recognised the potential of these specialist maps to strengthen his case for the wholesale sanitary reform of London, and it was during his involvement with the Metropolitan Commission of Sewers that he pushed his case most forcefully. Convened in December 1847, following recommendations made by a Royal Commission that had investigated sanitary problems in London,[6] the Metropolitan Commission of Sewers was a new, centralised governing body for the city's sanitation, superseding seven of the eight separate Sewer Commissions who had managed London's sewers since Henry VIII's Sewers Act in 1531.[7] From the outset the main charge of the new Commission of Sewers was to design and construct a unified drainage network for London. The report of the 1847 Royal Commission had built up a picture of a metropolitan sanitary system on the verge of collapse, a primary cause of disease, governed by a multitude of conflicting interests, and constructed piecemeal from information gleaned from local, parochial surveys.[8] The creation of the new Commission of Sewers was intended as a first step towards the improvement of London's sanitary infrastructure and the abolition of such existing defects.[9]

At the very first meeting of the Commission on 6 December 1847, Chadwick outlined his intentions for a new survey of London, requesting an application to be directed immediately to the Board of Ordnance to survey London at the unprecedented scale of five-feet to one-mile.[10] Chadwick viewed the creation of a large-scale accurate plan of the built-up area of London as an essential prerequisite to the planning of a citywide sewer system.[11] As early as 1842, he had proposed that the Board of Ordnance should carry out such a survey, making a strong case against the use of civil surveyors, whom he regarded as lacking the competence and discipline of the Ordnance Survey's soldiers. For Chadwick, the new mapping of London was to have a scientific and disinterested basis epitomised, in his view, by the Ordnance Survey.

The survey commenced on 26 January 1848 with the construction of observation posts in preparation for the triangulation of the city. A contingent of 250 men from the Corps of Royal Sappers and Miners – engaged at that time on large-scale surveys of towns in northern England – were brought to London to begin work.[12] Observation towers were constructed on top of Westminster Abbey and St Paul's Cathedral, on the summits of Primrose and other hills, on towers, steeples, and roofs of churches, and on the terraces of public buildings. Individual posts were visible from one another and from the main observatory atop St Paul's. The appearance of soldiers on the streets of London and the prominence of the observation posts attracted a great deal of public attention and, in some cases, alarm: some feared that the city was to be imminently invaded; others thought that the observation posts might be a 'gigantic machinery of espionage'.[13] The accounts in the *Illustrated London News* emphasised the pressing need for the survey and echoed Chadwick's intentions, laid out in his initial report:

> One would have thought that there had been surveys enough already; and so there have. The cost of many complete surveys has hitherto been expended in piecemeal irresponsible work, in which no public confidence could be placed ... and we trust that no narrow, interested views will be allowed to interfere with the same course of proceeding as to the great works of improvement of which this survey is the base.[14]

The notion of seeing London as a unified metropolitan area, governed by scientific and objective principles, runs through much of the press coverage of the survey.[15] What is consistently emphasised is the need for a new conception of the city – one that is not based on local vested interests, but rather unified and disinterested, seeing the city as a whole, undivided by the old parochial boundaries.[16] Existing forms of local government in London produced a profusion of boundaries: some 300 different bodies governed the city in the late 1840s, administering parishes and innumerable districts within those parishes.[17] The Ordnance Survey, as an external and supposed 'neutral' military organisation, ignored the confusion of jurisdictions created as a result of these many governing authorities; instead, it mapped London as a city without boundaries, even if this did not reflect the administrative reality at the time.

Contesting Views

The references to 'narrow, interested views' in the *Illustrated London News* account were indirect attacks on those who were opposing the new survey: civil surveyors who found a powerful representative in the Member of Parliament for Bodmin and prominent London mapmaker, James Wyld (1812–87). Wyld argued that the new survey of London was unnecessary; instead, he proposed that it should use existing plans put together by local surveyors at a far more reasonable cost.[18] Wyld eventually formed an alliance with the Association of Surveyors, a group that defended the interests of civil surveyors against the rising dominance of the Ordnance Survey; together, they proposed to commence their own survey of London with Wyld producing the map.

Chadwick was quick to respond: on 28 March 1848, the Commission met with Wyld to discuss his proposed plan. Henry Austin (*c.*1811–61) – consulting

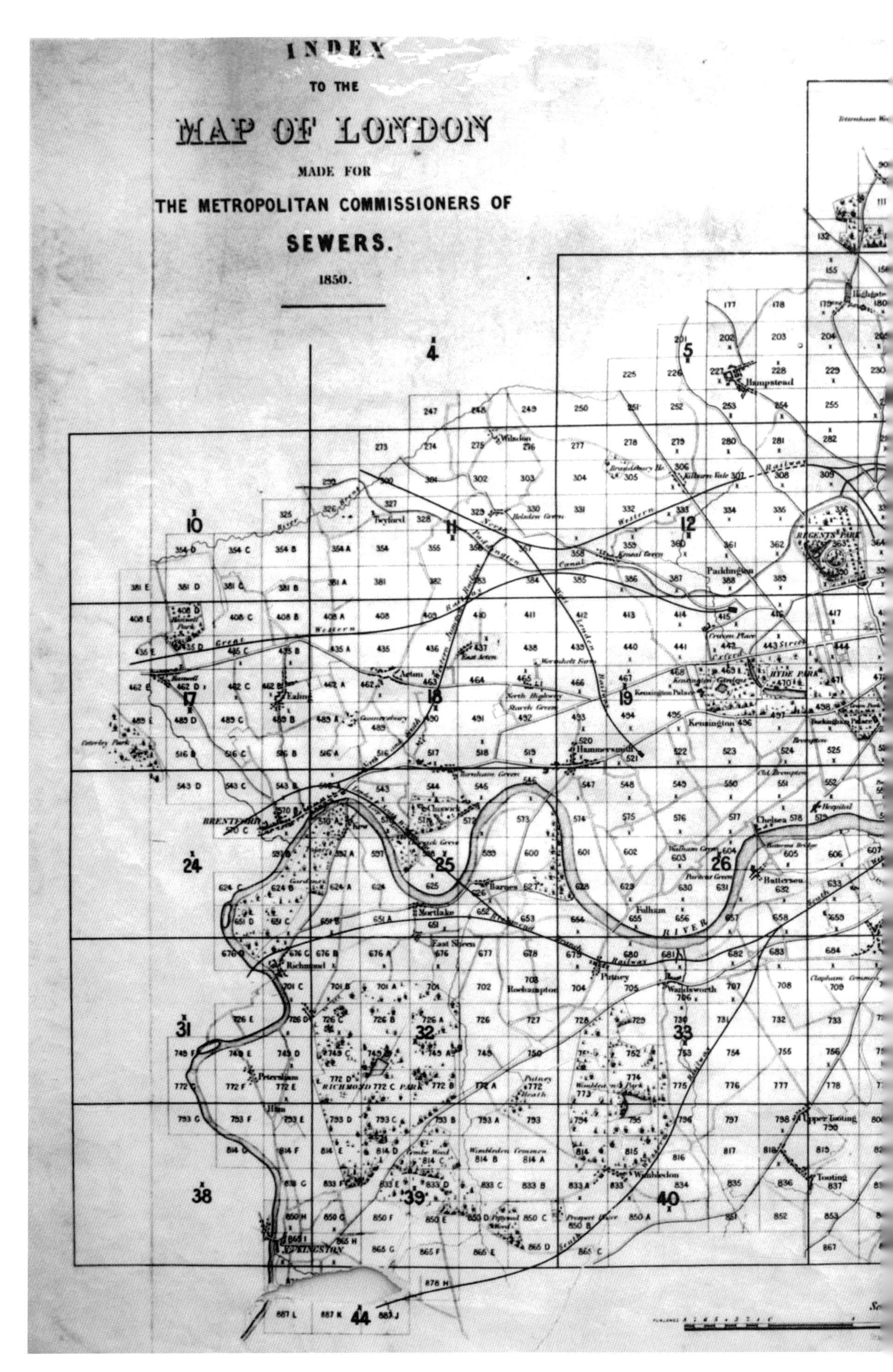

1.1 Index to the Map of London made for the Metropolitan Commissioners of Sewers produced in 1850 and showing the arrangement of the twelve-inch (larger rectangles) and five-feet maps (smaller rectangles).

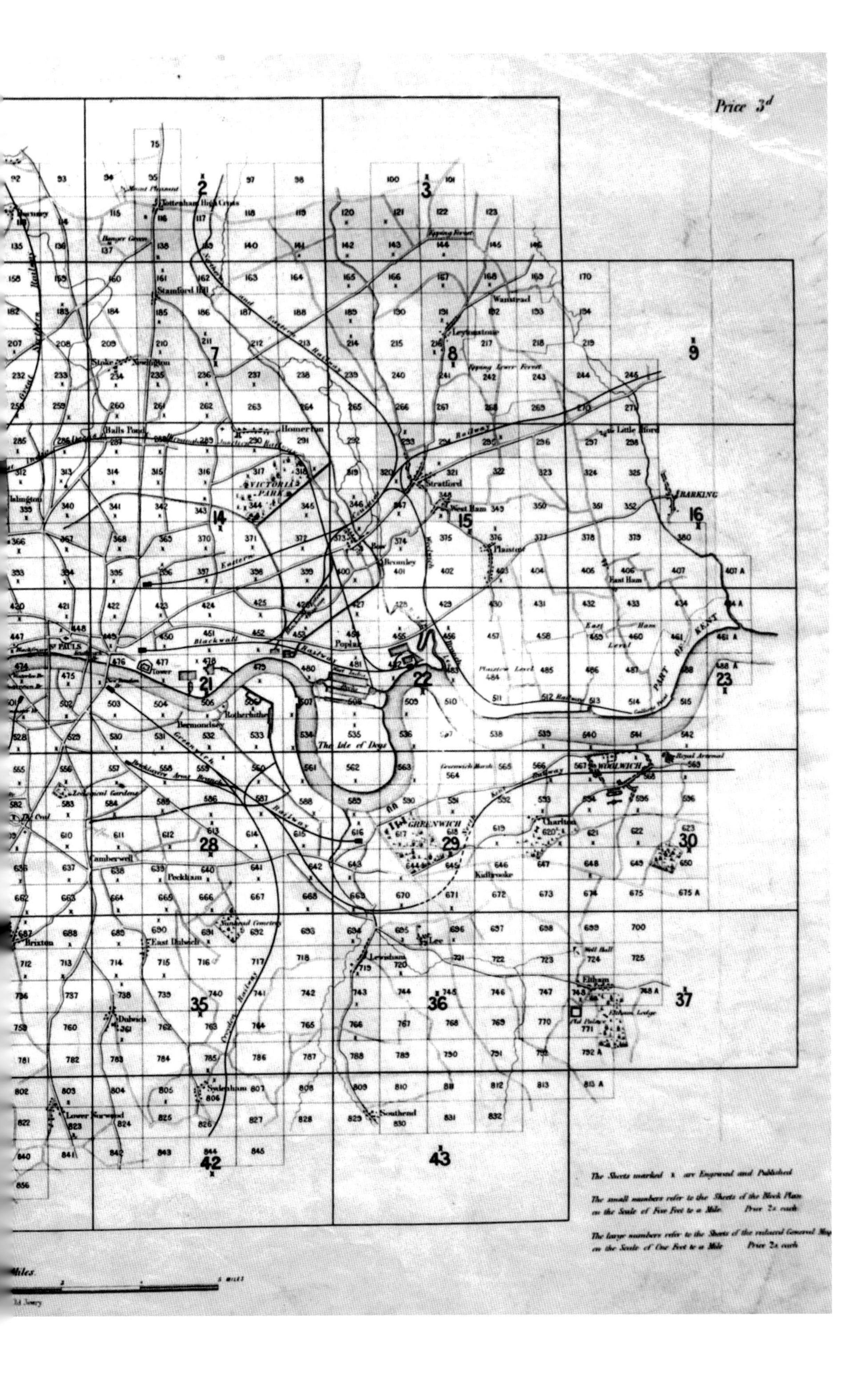
Price 3d
Tottenham High Cross
Stamford Hill
Epping Forest
Wanstead
Leytonstone
Stoke Newington
Balls Pond
Homerton
Victoria Park
Stratford
West Ham
Plaistow
Bromley
Little Ilford
Barking
East Ham
Islington
St. Pauls
Tower
Poplar
Blackwall
Plaistow Level
East Ham Level
Part of Kent
Bermondsey
Rotherhithe
The Isle of Dogs
Woolwich
Royal Arsenal
Greenwich Marsh
Greenwich
Charlton
Zoological Gardens
Camberwell
Peckham
Kidbrooke
Brixton
East Dulwich
Lee
Lewisham
Eltham
Eltham Lodge
Dulwich
Sydenham
Lower Norwood
Southend
The Sheets marked x are Engraved and Published
The small numbers refer to the Sheets of the Block Plan on the Scale of Five Feet to a Mile. Price 2s each
The large numbers refer to the Sheets of the reduced General Map on the Scale of One Foot to a Mile. Price 2s each
Miles

engineer to the Commission – took the opportunity to defend the need for a new survey of London. He admitted the existence of accurate large-scale survey plans of parish areas, many of which were produced as a result of the Parochial Assessment Act of 1836,[19] but stated that:

> ... the attempt to connect these different surveys into a whole would be utterly futile, ending only in failure .... They have been executed totally independent of each other, under different arrangement and different systems, and they never could be fitted accurately together. The slightest of discrepancy in the junction of one part would throw the whole work entirely out at another, and a work so put together would never possess the confidence of the public.[20]

With a primary emphasis on accuracy, Austin reiterated the need for any new maps of London to form an integrated whole, which he saw as missing from the older plans, such as those made by the Holborn and Finsbury Sewer Commission: plans directly inherited by the Metropolitan Commission when it was formed in 1847 (**Pl. I**). Represented on the map is a profusion of descriptive information: individual streets and buildings – copied from older topographic plans – form the base plan; overlaid are other descriptive features such as sewers (thin red lines) and a profusion of unidentified boundaries, some represented by thick yellow, pink and green lines, others by dashed black lines. All of this detailed information was specific to the area of jurisdiction of the old Holborn and Finsbury Commission and abruptly ends at the boundary of the district controlled by the Westminster Commission, who would have produced their own local maps according to their particular needs. Austin saw such isolated surveys as directly opposed to the notion of applying an accurate and consistent method of surveying to the city as a whole: that is, one necessary to make it possible to design a citywide sewerage system. As such, it was not only mapmaking procedures that were challenged by Austin and the Metropolitan Commission of Sewers, but also the entire existing political and administrative framework of the city.

The Metropolitan Commission of Sewers received Parliamentary validation for the completion of the new survey with the passing of the Metropolitan Sewers Act in September 1848.[21] Subsequently, the work was completed very quickly: the ground survey was finished by March 1849 with the measurement of road levels – information also shown on the completed maps – completed by May. By November 1849, 79 sheets of the five-feet survey had been engraved onto copper; by July 1850, the engraving of the 901 sheets that made up the complete survey was almost complete. The extraordinary speed of the survey – faster than even the original cost estimate had anticipated – resulted in an extra ten copies of the sheets being made (in addition to the ten already ordered) as well as the preparation of index plans at 12-inches to the mile, comprising 44 sheets, and a general index plan on a single sheet.

The View from Above

The general index plan (**1.1**) outlines the arrangements of both the five-feet to one-mile sheets (**1.2**), and the 12-inch to one-mile index sheets (**1.3**). The internal coherence of the individual sheets, at both scales, is emphasised in the general index plan (**1.1**); its grid-like form shows that the multitudinous sheets of the survey are definitely not pieced together in the manner of older plans; rather, they have their own internal structural unity – all of the sheets are shown to fit exactly together within the overarching framework of both the 12-inch (larger rectangles) and the five-feet sheets (smaller rectangles). This visual unity reflects the conception of the city propagated by the Metropolitan Commission of Sewers discussed previously: London was conceived of, and now represented as, a unified city.

Both the five-feet (**1.2**) and 12-inch sheets (**1.3**) present a clear promotion of this new image of the city. The existing parish boundaries of London are replaced only by the edges of the individual sheets, which fit together exactly to form a grid. Each of the 901 sheets of the five-feet map and the 44 sheets of the 12-inch map are clearly numbered; on the edges of each sheet are shown the numbers of the sheets to which they connected, giving the whole map a clear internal structural unity. The features depicted on both maps reveal other structural links: dots and numbers, scattered over the entire surface of the sheets (seen most clearly in **1.3**) mark the positions at which road levels were carefully measured, and small arrows indicate the location of the permanent benchmarks which allowed a correlation between this aerial view and observations at ground-level. These levels and benchmarks systematically linked all of the sheets together and also provided a foundation for calculating the gradients necessary for building a citywide sewer system. In short, the urban region had become stylised and comprehensible, in preparation for a future intervention – that is, the construction of a unified sewer system, made possible by the features depicted on these maps.

As abstract and totalising images, these maps embody what David Pike has described as 'the view from above' London – that is, panoramic views of the city, seen in both city maps and gigantic painted panoramas of the city, such as that displayed in the Colosseum after its opening in 1829.[22] The vantage point for city panoramas was the dome on top of St Paul's cathedral, which enabled observers to encompass the entire cityscape whilst still being able to pick out the smallest of details.[23] The Ordnance Survey, by establishing St Paul's as its central triangulation point, closely conforms to the typology of the panorama, albeit in a stripped-down format without its sense of perspective, visualising only those parts of the city that were related to the modernising agenda of the Metropolitan Commission of Sewers: namely, roads and waterways. Chadwick had clearly stated the reason for this: he wanted depicted only those features that were directly connected with the construction of a unified sewer system. The map can therefore be considered as a marking out of the territory needed by the Commission to construct this system. Roads and waterways are marked as accurately as possible, with variations in road width carefully measured and drawn, so that sewers could be accurately planned to run beneath them. This claimed territory also framed a space of future intervention: the spaces represented would be 'colonised' by the authority of the Commission as

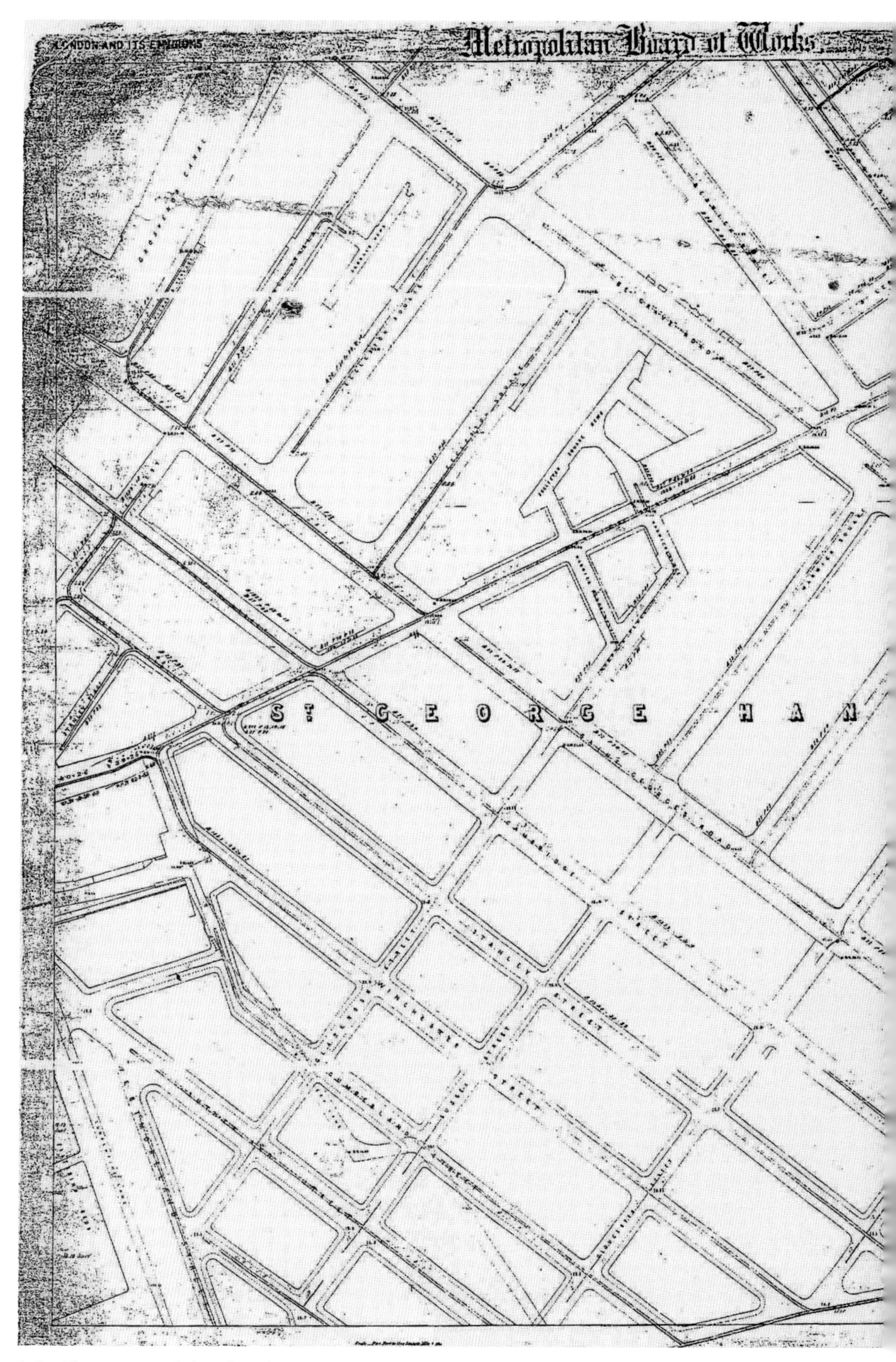

1.2 Sheet 553 of the five-feet to one-mile Ordnance Survey of London, *c.*1849. The lines seen in the centre of roads represent sewers – information added sometime in the 1850s and still used by Thames Water today.

526
ST MARGARET
WESTMINSTER
ST JOHN THE EVANGELIST
E R S Q.
580

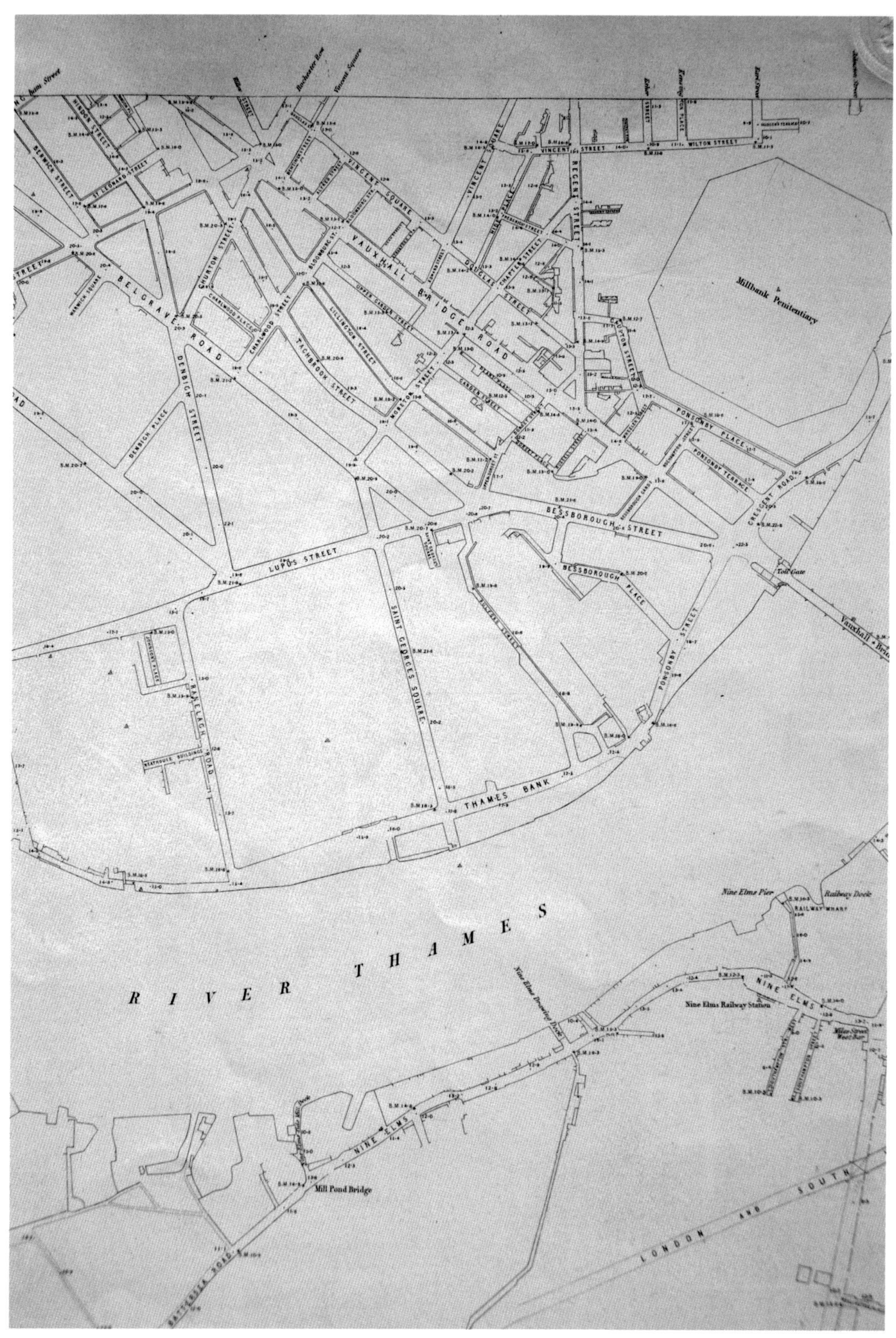

1.3 Extract from sheet 27 of the Ordnance Survey of London twelve-inch to one-mile skeleton plan, drawn up from 1848 to 1851 and printed by lithography.

they began to plan and engineer a citywide sewer system. Consequently, the Ordnance Survey maps can be read as highly subjective and 'interested' representations of the city, even if their producers claimed to have an objective interest in applying a universal, scientific method, as opposed to the subjectivity and narrowness of their opponents. Behind this claim of transparency from Chadwick and the Commission lies an implied ideology of improvement: that is, the desire to transform the city into a unified, clean and rational site, enacted and governed by themselves.

The View from Below

When Chadwick had originally outlined his plan for the proposed Ordnance Survey in January 1848, he also recommended that a subterranean survey be carried out of the built-up area of the city.[24] The subterranean survey had a very different goal from that of the Ordnance Survey: its eventual aim would be to produce maps depicting the fine details of the existing drainage, whether under the streets or in individual houses. Chadwick viewed the control of the old sewer spaces, with all their problems, as essential to the future success of any new system. This survey would involve a detailed investigation of the existing sewers within the area of jurisdiction of the Commission and the production of plans showing 'all courts and alleys ... [and] needful information connected with the drainage or absence of it'.[25] This literal descent into the 'bowels' of the city conforms to another common representational trope in Victorian London: the 'view from below'. Unlike the view from above, which placed the city's substratum at a safe distance, the view from below positively revelled in the dirt and chaos of subterranean spaces: in effect, a modern variant on the descent into the underworld seen in ancient mythologies.[26] However, for those who carried out the subterranean survey there was to be no such abandonment: it was to have the same scientific basis as the Ordnance Survey, in order to produce, in Chadwick's words, 'a true picture ... of the existing drainage beneath the surface'.[27] In effect, this view from below was to be as rigorous and panoramic as its counterpart from above: it was to bring the infernal regions within the compass of a new vision of rationality and unity seen in the Ordnance Survey. This aim raised a fundamental question: how to represent the old – already seen by Chadwick as standing in opposition to his vision of the new – using the same criteria that would define the new: that is, scientific rigour and precision? The maps and reports produced as a result of the subterranean survey were attempts to resolve this problem: they aimed to articulate a coherent relationship between the old and the new; between the view from above and the view from below.

On 13 April 1848 it was agreed that personnel from the Board of Ordnance – two levellers, two labourers, four other staff and two marksmen – be employed to assist with work on the subterranean survey, directed by the chief surveyor of the Metropolitan Commission of Sewers, Joseph Smith.[28] With such a tiny workforce – 250 men were used for the Ordnance Survey – the subterranean survey became a long and drawn-out process; in 1855 – seven years later – it was still unfinished.[29] The surveying methods employed were similar to those used in both surveys; however, they were applied in a very different working environment. Where the surveyors of the Ordnance Survey could build observation towers to view the entire city, the subterranean surveyors had to negotiate dark and claustrophobic spaces; where the public admired the gleaming uniforms of the military surveyors, those beneath the ground were hidden from public view and surrounded by filth.

The surveying of a sewer began with the determination of an access point – usually a manhole on the surface of a road or footway. Levellers would then descend into the sewer and lamps would be lit so that a theodolite could be used in the shadowy space to measure its level beneath the ground. The work was often dangerous, as many of the sewers contained deposits of foul matter several feet deep. Indeed, some of the men only narrowly escaped death by drowning or from explosions caused by sewer gases.[30] The levellers produced highly detailed but crude drawings in notebooks outlining the levels, dimensions and conditions of the sewers examined.[31] **1.4** is a reproduction of a page from one of these notebooks, the work having been carried out on 4 September 1848 by T. Bevan, one of the levellers employed by the Commission.[32] This drawing depicts part of the sewer in Lillington Street where it joined another in Upper Dorset Street, Westminster, both streets seen on the centre-right of sheet 553 of the five-feet scale map (**1.2**).

The reproduced drawing highlights the level of detail surveyed: the width of the sewer under Lillington Street is shown in the centre of the drawing with levels indicated in many places along its length. The rest of the sewer in Lillington Street is represented on the other pages of this notebook, with each page joining up precisely with the next. The point marked 'H' in the sewer in Upper Dorset Street indicates the place of entry into this section of sewer, where the surveyor would have descended with his measuring equipment. Every junction of the sewer is noted and represented: even the individual one-foot circular drains – leading from the houses flanking the road to the main sewer – are given a full description indicating their size, shape, and condition (here, described as 'clean'). The level of each junction above the invert (base) of the main sewer is shown, as is the cross-sectional shape of the sewer (seen just above the centre-left of the page), sketched roughly but precisely measured. The drawings in most of the other notebooks – crudely drawn in often-difficult conditions – replicate this same level of detail. Such voluminous records reflect Chadwick's intention to gain a comprehensive and minutely detailed picture of London's existing sewers.

Monstrous Spaces

In August and September 1848, Joseph Smith and Henry Austin presented written reports outlining the progress of the survey. Tables drawn up by Smith, which detailed the condition of sewers inspected, accompanied Austin's reports.[33] Both sections of the reports present an unremittingly bleak picture of London's existing sewers. Smith employs a vocabulary that gives emphasis to this picture: the recurring adjectives used are 'dilapidated', 'decayed', 'dangerous', 'defective', 'frightful', 'crushed', and 'rotten'. Smith even includes

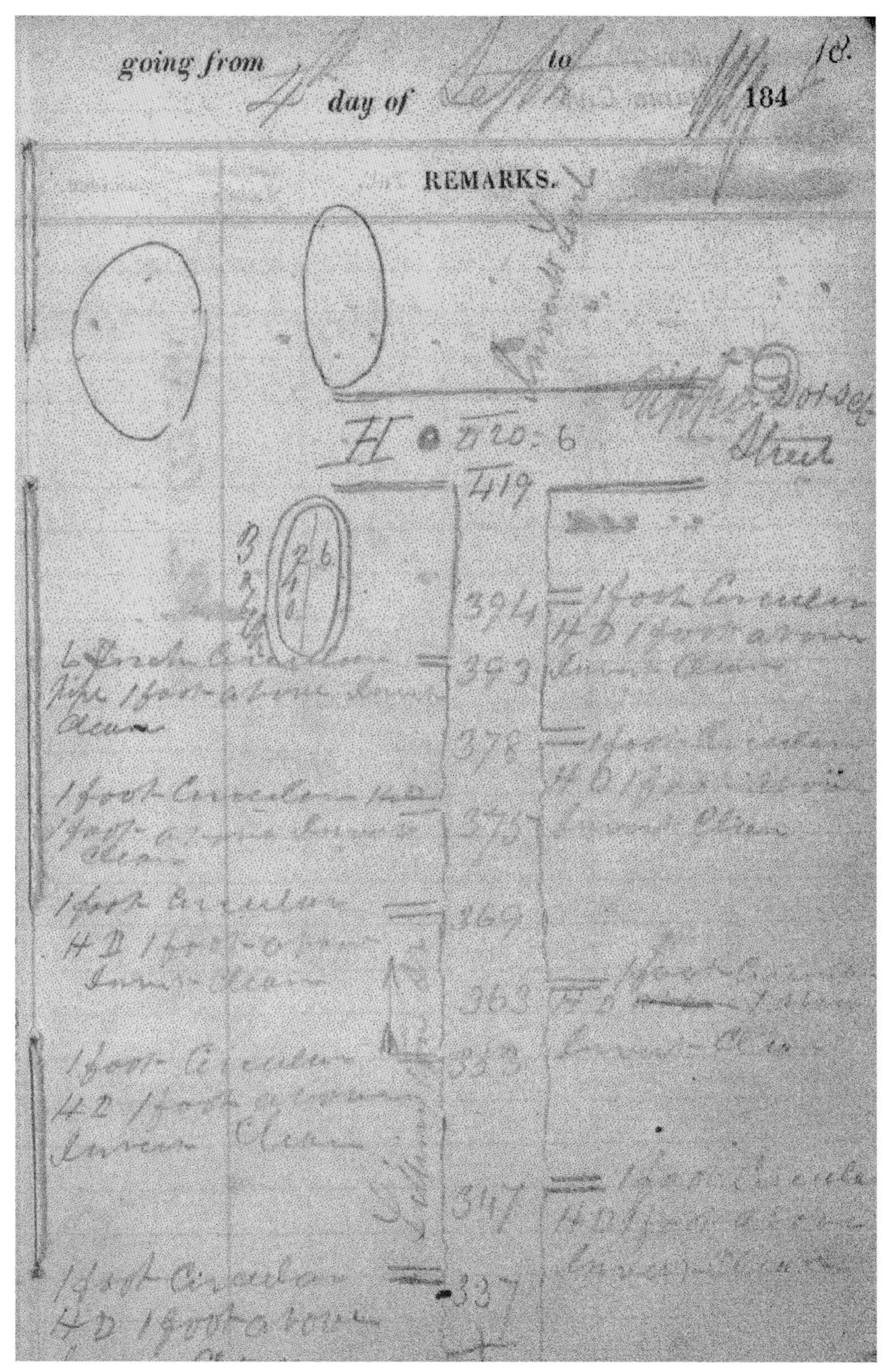
going from 4 day of Sept to 18.
184
REMARKS.

1.4 Westminster levels, book 45, p. 18, showing sewer information collected by the surveyor T. Bevan on 4 September 1848.

small diagrams in his tables to represent the poor state of particular sewers.[34] Austin's accounts emphasise the dangerous nature of the task of examining these sewers. Regarding the sewers in south London, he states:

> The smell is usually of the most horrible description, the air being so foul that explosion and choke damp are very frequent. On the 12th January we were very nearly losing a whole party by choke damp, the last man being dragged out on his back (through two feet of black foetid deposit) in a state of insensibility.[35]

In another incident – an explosion in the Kennington Road sewer – men working had 'the skin peeled off their faces and their hair singed'.[36] Such monstrous disfigurement contrasts graphically with the clean orderliness of the above-ground surveyors, noted in the press coverage of the Ordnance Survey. The below-ground survey, hidden from public view, is characterised by unexpected dangers, waiting to strike these purveyors of efficiency and order.

The dangerous nature of the work is closely related to the poor condition of the sewers, which are often depicted as irregular or badly-shaped spaces; when describing Blackland's sewer in the Chelsea district, Austin comments that it 'is egg-shaped with the broad end down .... The invert is of most irregular form'.[37] The defective shapes of the sewers led to the majority of them being blocked by excrement. Smith, referring to the sewers in Great and Little Chesterfield Street and Marylebone Court, comments that: '[t]hese sewers are of the very worst description, being so filled with deposit and ruinous material that portions of them cannot be perambulated'.[38] In addition to these blocked and dangerous sewer spaces, some were also in an advanced state of decay and dilapidation: Austin's report states that the sewer in Cumberland Street 'goes under the buildings of Upper Bryanston Street. The invert is all washed away, the side walls and arch falling in; if the sewer gives way the houses will follow';[39] another sewer in Westmoreland Street and Woodstock Street 'is totally decayed throughout the entire length. It is not safe, even for a day'.[40]

Whether sewers were decaying, dilapidated, blocked or irregular-shaped, all eventually became dangerous. Austin and Smith build up a picture of interrelated problems: a poorly built sewer, with its irregular spaces, led to it becoming 'choked with filth'; a poorly built sewer also displayed signs of dilapidation and decay which, in turn, led to it being classed as dangerous. Their observations consistently reinforce the notion of a disjointed series of sewers that urgently needed to be replaced.[41] This sense of disconnected sewers – built with no relation to each other – is highlighted in Austin's description of a junction in the Blackland's sewer: here '[t]he sewer in Cumberland Street joins it a foot below the invert'.[42] Like the old London maps – criticised by Austin in his report on the Ordnance Survey – these sewers did not join up correctly: they were 'pieced' together, which resulted in problems such as those seen in the Blackland's sewer where another sewer joined it a foot below the level of its floor.

These reports give an added interpretation to the information collated by the surveyors in their notebooks, relaying a sense of the heroic in the carrying out of the subterranean survey; they chart with an unflinching eye (and nose) these unknown and dangerous spaces beneath the city in all their horror. However, the notebooks, free of any descriptive language, contain series of interconnected drawings and tables, carefully pieced together from multitudinous sections – in effect a rationalised and ordered view from below. Such differing representations highlighted the problem of how to bring together the view from above and the view from below; how to represent the old – which, by definition, stood in the way of the new – in a way that conformed to the 'objective' methods employed in the Ordnance Survey. In their disconnected and dirty state, the old sewers defied the new vision of the city depicted in the Ordnance Survey plans – that of a unified series of interconnecting spaces. Smith and Austin censure the existing sewer spaces with appropriate and interrelated language, translating the rigorous and scientific work of the surveyors into a condemnatory picture and potent testament to the urgent need for a new unified system of sewers. However, by employing such 'loaded' language, they eschew the sense of objectivity that the Metropolitan Commission of Sewers claimed the Ordnance Survey embodied.

Combining Views

On 21 March 1849, the Commission met to discuss the progress of the subterranean survey and how to map the information gleaned.[43] They decided to trial a new map at double the scale of the Ordnance Survey, that is, an astonishing ten-feet to one-mile. They argued that the five-feet survey was not at a sufficiently large scale to include details of house drainage – details that Chadwick viewed as essential to understand and control before any new system could be properly constructed. The Commission resolved to start with the area covered by sheet 527 of the five-feet map to test the viability of producing further maps at the larger scale. Austin presented survey drawings in relation to this sheet on 20 April and, by June, the scope of the ten-feet survey had been extended to cover 'special districts and portions of districts'.[44] The areas eventually mapped were the City of Westminster and an area of Kensington near to Holland House, with 28 sheets being produced in total – 13 for Westminster and 15 for Kensington.[45] Although the production of new maps at a larger scale was experimental, the districts singled out for immediate survey were those considered to contain sewers and household drains that most urgently required repair or rebuilding.

Fig **1.5** shows an extract from one of the sheets of the ten-feet to one-mile survey, encompassing the area shown in the surveyor's notebook (**1.4**) – the junction of Lillington and Upper Dorset Street is shown at the centre right of this extract. This map is closely linked with the data collected in the notebooks of the surveyors; their detailed information was to be superimposed onto these large-scale plans. However, not one of the original sheets made at the larger scale actually depict house drains; at some point the Commission abandoned the notion of mapping drainage at this scale.[46] The maps produced on the ten-feet scale were hand-coloured and, apart from one experimental sheet, were never engraved.[47] Reproduced on the extract shown in **1.5** is similar information included on the Ordnance Survey maps (**1.2** and **1.3**), such as levels and benchmarks. However, the range of detail added is unprecedented. Particular attention is paid to the layout of individual properties: the map shows the boundaries of all houses as well as their interior sanitary arrangements –

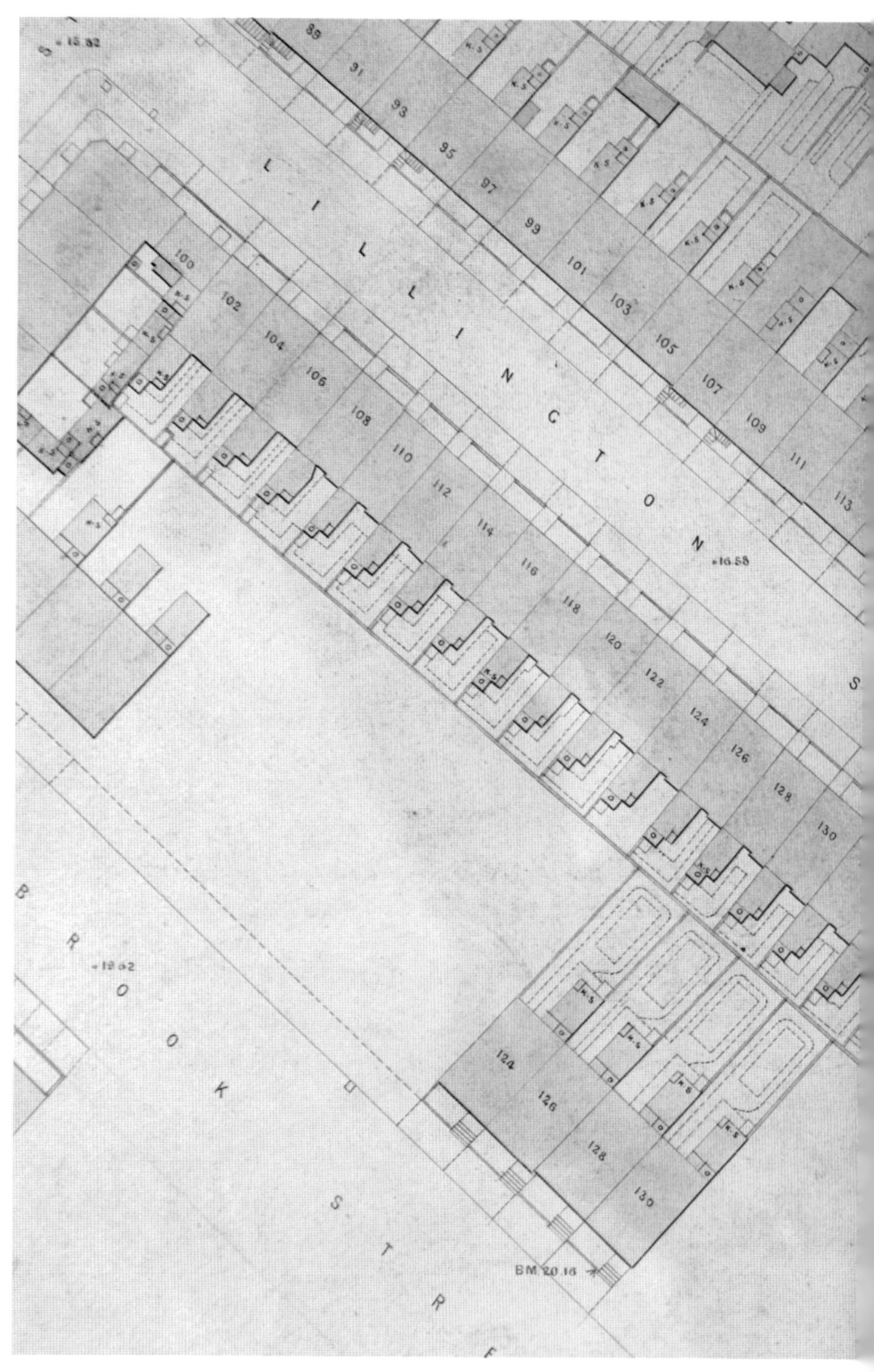

1.5 Extract from a plan of Westminster at the unprecedented scale of ten-feet to one-mile, referring to part of sheet 553 (SE) of the Ordnance Survey of London five-feet to one-mile maps, *c.*1850.

GOODMANS COTTAGES
ELDAR PLACE
DORSET STREET
DORSET PLACE
STREET
BM 12.51
BM 17.19

small circles indicating where flushing toilets or cesspools were situated. Even specific types of paving are shown on parts of the map, suggesting that this would have been extended to cover the entire sheet.[48] In addition, a colour scheme was proposed to distinguish between sewers that were sound, decayed, or in danger of collapse.[49] However, this map, like the rest of the ten-feet plans, remained as a base plan with none of the details of house drainage added.

Pl. II shows the level of detail that was eventually adopted by the Commission in 1850 after the ten-feet sheets were abandoned. This extract shows sewer lines (coloured red) hand-drawn onto sheet 27 of the 12-inch to one-mile Ordnance Survey map (**1.3**).[50] In this map, the representation of the old sewers – seen in Austin and Smith's reports and the surveyor's notebook – is dramatically transformed: gone is the detailed information so rigorously collated by the surveyors; and gone is the sense of these sewers as disconnected and disjointed spaces – characteristics that dominated Austin and Smith's reports. Depicted instead is an obvious network of sewers seen over a wide area with major sewers, such as the King's Scholars Pond Sewer (shown at the top left of the sheet), given visual prominence by large-scale labelling. The interrelation of the existing sewers is stressed rather than specific indicators of their condition. Additional information is also overlaid onto the base plan: parish boundaries (black dashed lines); the names of some of the individual parishes (St John the Evangel[ist] at the top centre and right-hand side); and waterways (coloured blue). These features, omitted from the skeleton plan (**1.3**), serve to introduce descriptive elements that highlight the existing topographical features and boundaries of the city. In short, they transform the skeletal plan into something quite different. If the original intentions of the subterranean survey – to gain 'a true picture … of the existing drainage beneath the surface' – are diluted by this map then the structural unity of the Ordnance Survey is also undermined by the superimposition of a representation of the old on top of a potential new. In short, this map is no longer a unified view of the city, but an uneasy combination of opposing elements: old and new; decorative and abstract; divisions and unity.

This combination map also represents an attempt to combine the two views of the city described above: the view from above and the view from below. With this conflation a new a play of meaning is put in motion: the view from below, with its old, dirty and decaying sewer spaces and confusion of administrative boundaries disturbs the abstract unity of the Ordnance Survey plans – the view from above. Rather, this map suggests an incomplete and more complex meaning. The depiction of the old sewers, built under the control of the old Sewer Commissions, and the reintroduction of the existing parish boundaries that determined the limits of this control, undermine the structural unity seen in the Ordnance Survey plans. The city can no longer be seen as a seamless whole; instead, the existing and the old impede the assertion of such a 'totalising' view, bringing out the contradictions at the heart of Chadwick's ideology of improvement. The scientific rationale of the Ordnance Survey, which mirrored Chadwick's own new vision of the city, could not quite incorporate the old spaces into its dominating framework. For a moment, the two coexisted in a state of unresolved tension. What I go on to show in the next chapter is that this moment was relatively short-lived: during the 1850s, a new vision emerged for London's sewers, one that abandoned any attempt to map the old and instead focused its attention squarely on the new.

2

Sewer Space and Circulation

When we flush our toilets or empty our baths and sinks, we start a flow of sewage from our homes to the sewers in the streets and then onwards to the nearest treatment works. With a continuous supply of water, we take for granted this ceaseless circulation of wastes, something that simply does not exist in many other countries of the world. This principle of circulation – the uninterrupted movement of things in space – originated in the Renaissance. From then on, health, whether manifested in the human body or in the 'body' of the city, was defined by the principle of motion: 'the city in its very design [was] to function like a healthy body, freely flowing'.[1] This circulatory ideal not only guided the later development of urban infrastructure, whether water supply, waste disposal or transport networks, but also the motion of goods, money and people in the city.

This chapter considers the importance of ideas of circulation in the planning of a new sewerage system for London in the 1850s, concentrating on two dominant but largely divergent conceptions: first, Chadwick's idealised notion of 'cosmic' circulation, which influenced many engineers, particularly Henry Austin; and second, the engineers Frank Forster (1800-52) and Joseph Bazalgette's rationalised conception of flow in the development of their intercepting scheme.[2] The eventual construction of Bazalgette's system, after 1858, demonstrated how his idea of circulation had triumphed over that of Chadwick and his engineers. As in the re-mapping of London examined the previous chapter, visual representations played a key role in the promotion of opposing conceptions of sewer space and circulation. They allowed complex technical arguments to be clearly visualised in a comprehensible format, were vital in the persuading and convincing of others of their validity, and, in the end, played an important role in the eventual triumph of Bazalgette's plan.

Cosmic Circulation

At the opening meeting of the Metropolitan Commission of Sewers on 6 December 1847, Chadwick focused the attention of the Commission on the need for a new survey of the city. Yet, he also had another pressing concern: that the existing sewers in London be flushed and cleansed 'immediately'.[3] As

Chadwick made clear, both of these actions were essential prerequisites for a new sewerage system for London – a system that had been evolving in his mind for many years. In his 1842 *Report on the Sanitary Condition of the Labouring Population of Great Britain*, Chadwick had focused on the relationship between disease and poor sanitation and had emphasised the necessity of providing new drains for houses to remove disease-ridden decomposing matter.[4] Chadwick regarded the imposition of a centralised form of administration as key to the development of improved sewerage and water systems.[5] It was during his involvement in the Health of Towns Commission in 1844 that he developed more fully his concept of an urban 'system' of sewers, connecting every household to improved street sewers, cleansed by an improved water supply that would quickly remove any accumulated deposits.[6]

The evolution of Chadwick's technological vision was guided by one overarching principle common to many mid-Victorian sanitary projects: the hope that 'the agricultural use of urban sewage could finance much urban improvement'.[7] Chadwick's determination to use human wastes as agricultural manure reflected a growing interest in Britain in the 1840s regarding the possibility of turning waste into profit,[8] and formed the subject of many Parliamentary committees and debates in the second half of the 19th century.[9] Chadwick's ideas on sewage utilisation, or recycling, were derived from the German chemist Justus von Liebig (1803-73), who was one of the leading promoters of sewage recycling as an essential requirement for the long-term sustainability of agricultural productivity. Liebig thought that if urban wastes were not restored to the countryside the soil would be eventually drained of all its fertilising properties. For him, pouring sewage into the sea was literally squandering the precious reserves of the country[10] and he denounced England as 'the most flagrant practitioner of such profligate behaviour'.[11] The French novelist Victor Hugo, in his 1862 novel *Les Misérables*, gave a dramatic positive alternative to this vision of doom. Referring to the sewage of Paris he states:

> Do you know what all this is – the heaps of muck piled up on the streets during the night, the scavengers' carts and the foetid flow of sludge that the pavement hides from you? It is the flowering meadow, green grass, marjoram and thyme and sage, the lowing of contented cattle in the evening, the scented hay and the golden wheat, the bread on your table and the warm blood in your veins – health and joy and life. Such is the purpose of that mystery of creation which is transformation on earth and transfiguration in Heaven.[12]

Chadwick's own conception of sewage recycling was no less dramatic than Hugo's 'cosmic' vision of the full integration of human wastes into the natural cycle. In 1845, he stated that the overriding goal in any sewerage system was to 'complete the circle and realize the Egyptian type of eternity by bringing as it were the serpent's tail into the serpent's mouth'.[13] Such a dreamlike image envisaged a universal type of circulation where, with the continuous injection of urban sewage into farmers' fields, a type of agricultural productivity that could only be dreamed of would be sustained for an indefinite period of time.

This cosmic ideal represented a peculiar bringing together of economics and theology. On the one hand, sewage recycling made economic sense: as Liebig argued, it would release a previously untapped 'mine of gold' and simultaneously sustain rapid population growth in urban areas and the agricultural productivity necessary to sustain that growth. And yet, sewage recycling was also a solution to an important theological dilemma: was human waste really part of God's bountiful creation?[14] In the mid-19th century, natural theology made commonplace the notion that the character of God was self-evident in the laws of nature. Therefore, for Chadwick and many others, waste and decay were seen as 'a perversion of natural cycles which occurred only when organic wastes were not quickly returned to the soil'.[15] The almost obsessive interest in sewage recycling in this period was one way of resolving the theological dilemma raised by the idea of human waste itself. It is no wonder then that Chadwick's notion of cosmic circulation was expressed in metaphorical imagery similar to that found in a theological treatise; it represents both the realisation of an economic dream and also obedience to a divine imperative.

From this cosmic ideal, Chadwick developed his technological vision. He viewed liquid sewage as the best form of manure and consequently argued for a method of irrigation to be used – that is 'the application of liquid sewage by spray or ditch to arable land'.[16] Consequently, sewage needed to be fresh and dilute and therefore required rapid removal from urban centres to outfalls and then on to farmers' fields: optimal sewers were thus velocity-augmenting egg-shaped pipes; outfalls needed to be located close to agricultural areas; and rigid controls needed to be enforced as to what was or was not allowed to enter the sewers. Chadwick's technological vision was given its most coherent expression in an 1852 report by the General Board of Health, which he joined in 1848 after being expelled from the Metropolitan Commission of Sewers due to an intractable personality clash with fellow commissioner John Leslie.[17] In this report, which laid down principles for the design of sewerage systems for any town or city in the country, Chadwick's notion of a sewerage system evolves from the details of house and street drainage, commencing with the removal of any 'viciously-constructed drain'.[18] As also emphasised in chapter 1, Chadwick's condemnatory attitude towards London's existing sewers reinforces and validates his notion of creating a system *de novo*, free from the old, disjointed sewer spaces and the fragmentary legislation that governed them. In place of the old sewer spaces, Chadwick proposes:

> A tubular system of drainage, in combination with [a] constant water service [having] no decomposing deposit, no evaporating deposit, which is appreciable in house-drains or sewers.[19]

Chadwick's 'tubular system' consists of small earthenware pipes, as opposed to the existing large brick sewers, and their size and inclination is determined by the overriding need to maintain a rapid flow. His ideal is that all household sewage be 'immediately received in water, and carried along with considerable rapidity … before decomposition can have advanced'.[20] Governing all of these design features is the recycling imperative: to achieve a completion of the natural cycle all the distinct parts of the system needed to promote a constantly moving stream from house to street, from street to outfall, and finally, from outfall to fields.

Ideal Bodies

Chadwick's understanding of sewage circulation, to some extent, was based on the model of the human body. Indeed, commentators on Chadwick have described his tubular system as 'arterial-venous' – that is, directly modelled on the internal workings of the body: the arteries representing the sewers, the veins water pipes.[21] If the sewers and water pipes envisaged by Chadwick were technological interventions they were also 'organic' in that they coupled themselves directly and literally to the 'vital economy of the body'.[22] One consequence of Chadwick's organic metaphor linking body and city was an increasing emphasis on strict regulation, focused on the potential 'users' of his sewers: the citizens of the city.

It is not surprising, therefore, that Chadwick's technological vision focused on the prevention of the possibility of blockages in his system, especially in the domestic sphere. If the sewer spaces beneath the city streets were a vital part of his circulatory ideal, so were the sanitary fittings in the homes connected to that system:

> Arrangements should be made for cleansing the improved privies, urinals, and sinks in lodging houses, by means of jets of water instead of the broom. For this mode of rapid and complete cleansing, earthenware will be highly convenient.[23]

Chadwick views the use of a broom as an improper method of cleaning – it might sweep potential blockages into the sewers; only clean jets of water are allowed, flushing away deposit instantly. The system would also remove only household sewage; separate sewers would carry away rainfall, liable to collect road grit that might block the system.[24] For Chadwick, this vision of a perpetual flow in the sewers depends not only on active and rigorous inspection but also on the exclusion of certain bodily practices on the part of the user: in effect, strict regulation extends from the street to the house to the very body of the householder. Chadwick proposes that:

> No drain whatever should be allowed to be without a perfect protection to every opening, that the protections should be so secured and arranged as to be immovable, and the drains rendered inaccessible for surreptitious and improper practices.[25]

Once 'perfect' protection had been secured over any external variables, including the 'improper' practices of the user (which remain curiously undefined by Chadwick), a new network of pipes could be conceived, leading outwards from house to street to main sewer to outfall. Chadwick envisaged a new improved and centralised water supply connected to individual rooms in

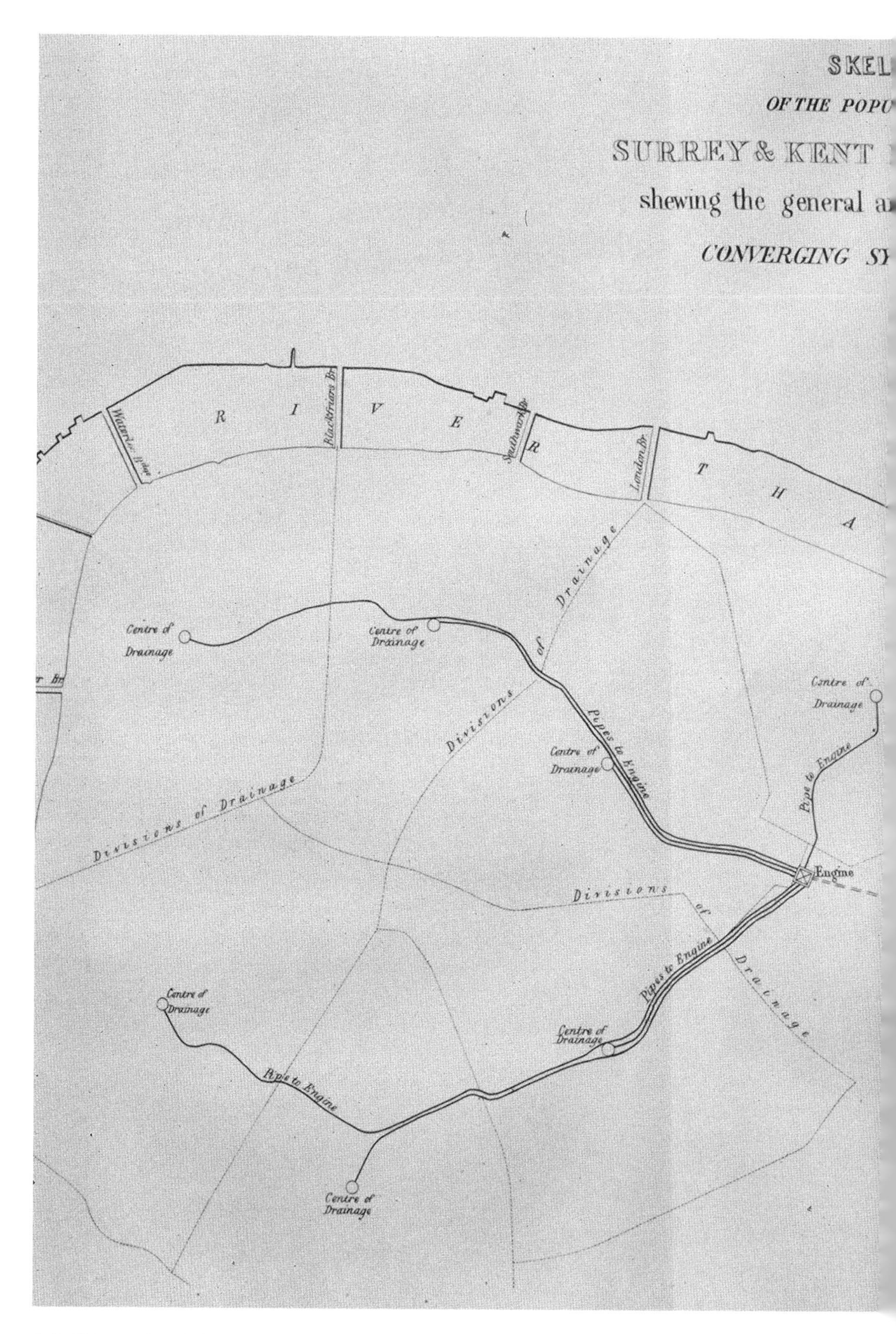

2.1 Henry Austin's proposed converging system of drainage for south London published in the first report of the Metropolitan Commission of Sewers in 1848.

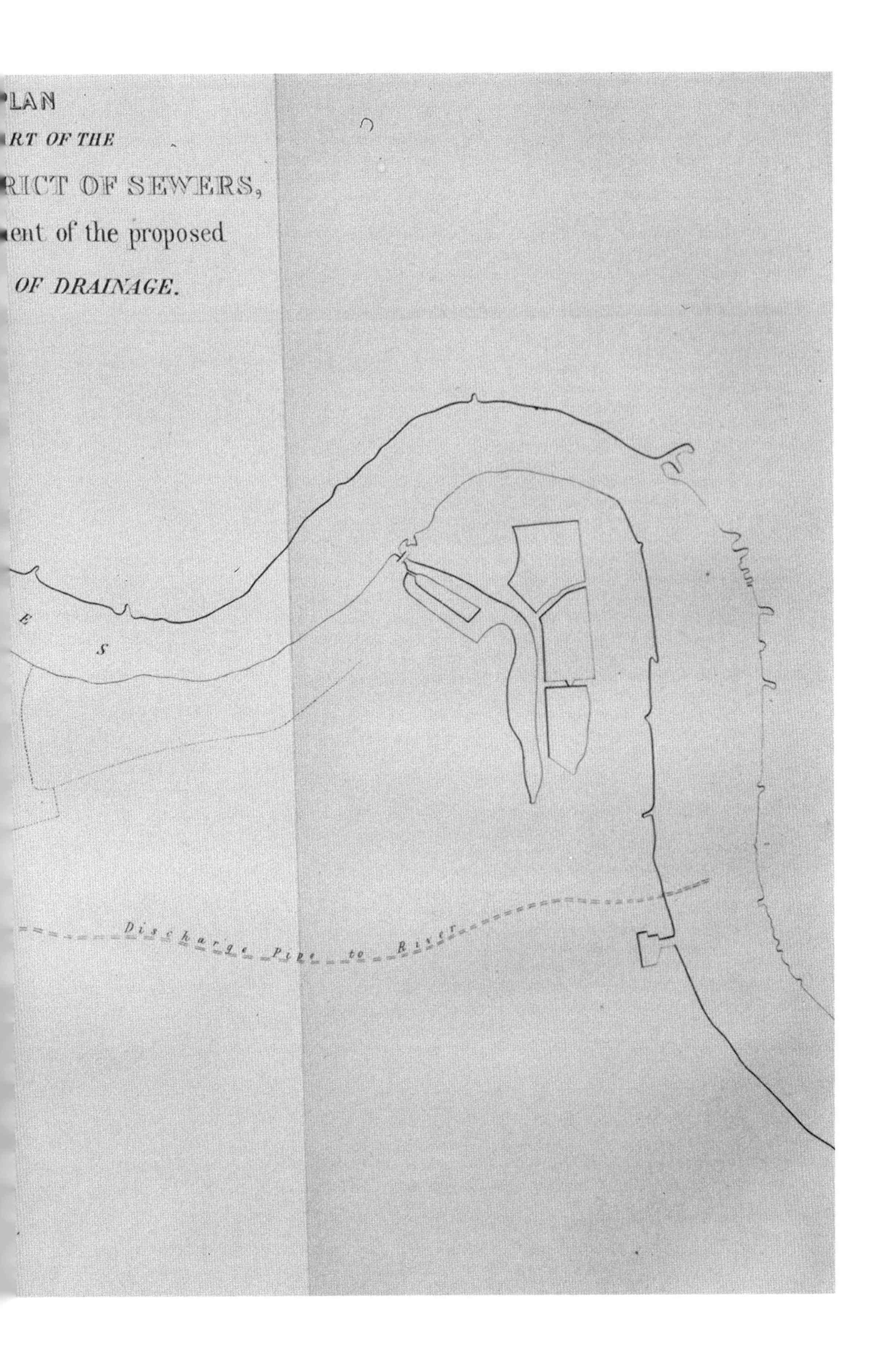
LAN
RT OF THE
RICT OF SEWERS,
ent of the proposed
OF DRAINAGE.
E
S
Discharge Pipe to River

blocks of flats, and also develops the concept of 'back' drainage, where every house would be equipped with individual drains leading to a common sewer running at the back of terraces.[26] This notion of 'communal' drainage reflected the paramount importance of control in this system: communal drainage, with one common sewer, is easier to inspect and monitor than individual house drains, and it reduces the possibility of interference from the user who is most certainly not trusted by Chadwick.

If Chadwick defined his technological vision by the notion of ideal circulation, he also insisted that this vision could be applied to every town and city in the country. He conceived of the 1852 report as setting out the universal principles of urban sewerage systems to be applied by local Boards of Health.[27] This aspiration to universality highlighted a fundamental problem at the heart of Chadwick's vision: how to apply general principles to specific towns and cities with their individual topographic characteristics? Chadwick's insistence on sewage recycling also came before any market for sewage manures had been established and consequently the location of the outfalls of any system remained unspecified. It is not surprising, therefore, that his vision tended to concentrate on the details of house drainage and the transformation and control of existing 'defective' sewer spaces and sanitary practices. Ideal circulation would not easily translate to the specific problems encountered in individual towns or cities: indeed, its very claim to universality would undoubtedly be undermined when any specific problems were encountered.

Ideal Spaces

During his involvement with the Metropolitan Commission of Sewers, Chadwick sought engineering expertise to turn his vision into a viable scheme for London. The Commission's consulting engineer, Henry Austin, developed a plan along Chadwickian lines for the drainage of south London (or the Surrey and Kent districts) that was intended as a preliminary plan for the entire metropolitan area. Compared to the north side of the River, this area was relatively undeveloped – the low-lying land subject to frequent flooding. Austin disparages the state of the existing sewers with their 'sluggish currents' creating 'foul air' from 'want of fall'. Instead, he proposes an entirely new system of sewers:

> The district ... should be apportioned into convenient sections or divisions, the drainage of which would be totally independent and distinct, converging to the centre of each division with any desired current, and from these centres the liquid would be raised by steam-engines, placed at any convenient point in connexion with them by pipes.[28]

Austin does not specify his proposal; rather, by using the word 'convenient', he maintains the possibility of an open-ended application of his scheme, with the 'divisions' of the district and currents in the drains remaining shapeless and indefinite. Austin also included a map with his proposal, which he states 'explains his meaning' (**2.1**).[29] However, this map, rather than grounding Austin's proposal in the specific topography of south London, instead serves to emphasise its essential lack of specificity. Shown on the map are the proposed divisions of drainage; the centre points, where house drains would converge; and the point marked 'Engine' where the collected sewage would be pumped by steam-engines to farmers' fields. From this pumping point Austin envisaged, at some point in the future, the laying down of 'distributing pipes from the engines in the direction of demand'. According to Austin, this 'direction of demand' could not be specified because 'the public mind had not yet been brought to appreciate the value of this material, and to apply it to its legitimate purpose, instead of throwing it away'.[30] Recycling remains a speculative element within this plan and because it was a primary and overriding goal the other parts of the plan (the divisions of drainage and proposed lines of sewers) could not be specified. However, despite this, the 'cosmic' ideal of circulation remains the overall governing vision – a vision that Austin never abandoned.[31] Rather than develop a site-specific proposal for south London, Austin, with his sketchy plan and indefinite descriptions, is unable to localise Chadwick's universal principles. As Austin makes clear, this scheme is simultaneously a plan for south London and also a universal system 'alike applicable to a village or a metropolis, the one being only a multiplication of the other'.[32]

Following Chadwick, Austin's report concentrates on the details of house and street drainage and includes a diagram indicating the arrangement of the smaller house and street drains leading to the centre wells (**2.2**). Starting with what Austin regards as the smallest possible size for a drain – nine-inch diameter – in the smaller streets (indicated on the peripheries of the diagram), he envisages a progressive increase in the sizes of the pipes – to a maximum of 24-inch diameter – leading to the centre wells. From these wells, collected sewage would be discharged to the pumping engines by means of stoneware pipes. Austin's depiction of a 'network' of pipes in **2.2** is schematic and conceptual; the grid-like arrangement presents a universal picture of an urban network of pipes that in reality would have to be grounded in the irregular arrangement of streets within the city. Such an idealised network expresses Austin's desire for a 'complete system of pipe drainage' that, without any specific topographical engagement, could only be expressed in this schematic and idealised form.

Austin's idealised illustrations are accompanied in his report by an emphasis on the rigid control of present insanitary domestic practices, mirroring Chadwick's own concerns. In Austin's scheme:

> Yard drains would not be left unprotected; sink gratings would be effectively secured; surreptitious openings would be impossible; and that form of water closet basin which will admit of such an abuse, either from the carelessness of servants or the mischief of children, would at once be abolished.[33]

The lack of specificity regarding the positions of the sewers in Austin's report contrasts sharply with the extreme detail given above: like Chadwick, Austin emphasises the creation of 'a proper regulated system', with strict controls being placed on any 'abuse' that might impede circulation in the sewers.[34] Mirroring the language of the reports he produced for the subterranean survey

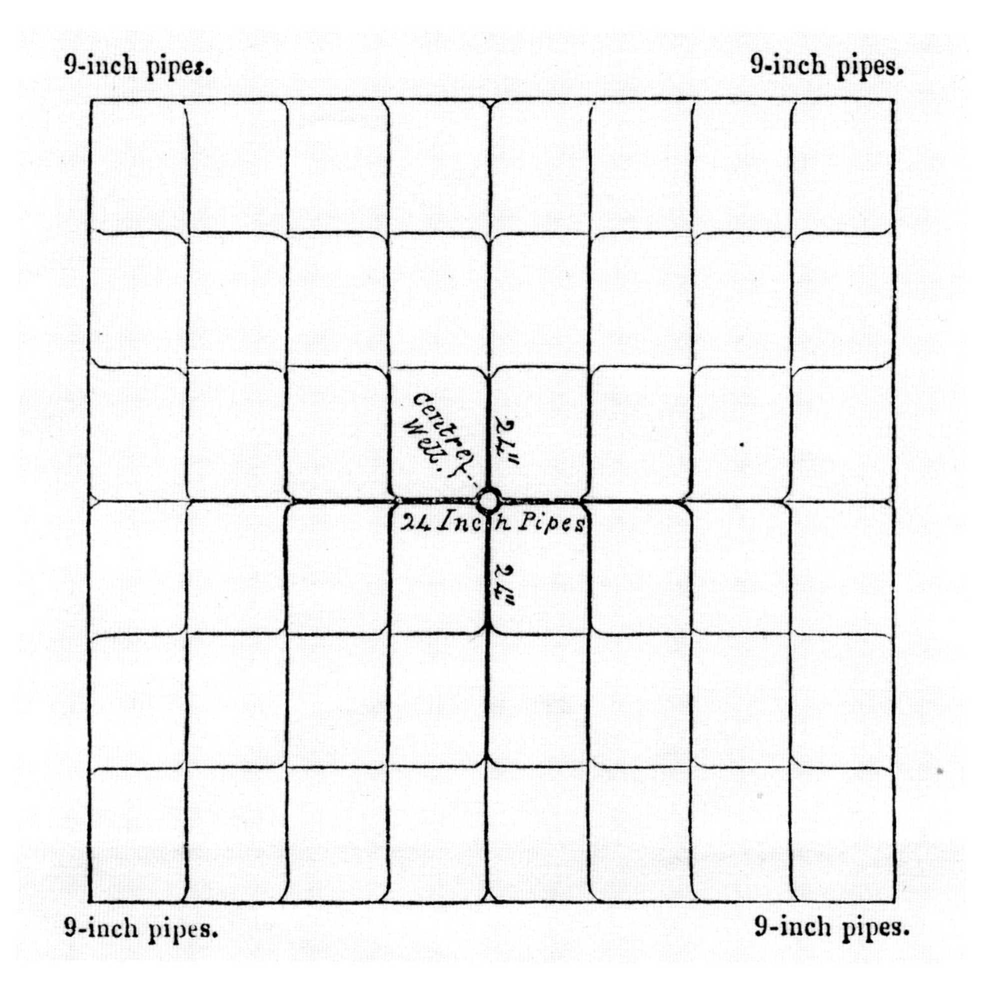

2.2 Henry Austin's diagram of a sewer network intended as part of his proposed converging system of drainage for south London, published in 1848.

in 1848 (see chapter 1), Austin condemns the present state of existing sewers:

> Perfection never can be attained until these brick house-drains are removed, and I believe it to be one of the greatest advantages of the converging plan of drainage proposed, that the whole system may be so reduced in size as to abolish altogether the use of brickwork even in the sewers.[35]

For Austin, brick is equated with the old sewer spaces, which he sees as unnecessarily large, poorly constructed and liable to blocking. Instead, he proposes the construction of new, small earthenware pipes and, like Chadwick, favours an egg-shaped cross section to increase the velocity of flow. As discussed in chapter 1, condemnation of the old sewer spaces thus becomes an important legitimating factor in the promotion of a new system. At the end of his report, Austin sums up with an overview of current sanitary evils:

> Imagine a baker, with a foul cesspool in his yard in close company with a well, and a choked drain in his kitchen – a combination by no means difficult to realize; the very heat of his oven brings a stream of poisonous

> atmosphere into his bakehouse, adding further contamination to every loaf that he has made with the already poisoned water from the well behind.[36]

Such an encyclopaedic description of sanitary spatial evils, characterised by overcrowding and blockages, is suggestive of a very different type of circulation: in effect, a chain of contamination. This picture of existing conditions sums up all that is detached from the new circulatory vision and, as such, is subversive and threatening. All that chokes and blocks is viewed as unhealthy; the new system would be characterised by the perpetual motion of water and wastes and the elimination of anything that threatened to block this motion.

Rational Spaces

Chadwick's departure from the Metropolitan Commission of Sewers in 1849 was followed almost immediately by the resignation of Austin as its consulting engineer.[37] At that time, the commissioners were unable to reach agreement as to the best plan to adopt and it was decided that an open competition should be held so that new plans could be submitted, and an invitation for tenders was issued on the 23 July 1849.[38] On 16 August 1849, Joseph Bazalgette was appointed, for one year, as an assistant surveyor at an annual salary of £250.[39] Bazalgette's appointment marked the start of his involvement with the Commission that would last until its demise in 1855. In his early years, Bazalgette had worked on land drainage projects, specialising in the reclamation of marshlands. In 1842, at the age of 23, he set up his own engineering practice and participated in the railway boom of that period. Like many of his colleagues in the profession he worked himself to a breakdown and in 1847 he retired to the country to recuperate.[40] In 1849 he returned to London and applied for the post of assistant surveyor to the Metropolitan Commission of Sewers, to which he was appointed in March.

One of Bazalgette's first tasks, with help of Edward Cresy (another assistant surveyor), was to examine the 137 plans tendered as a result of the competition, submitted by a mixed array of surveyors, engineers and amateur enthusiasts. Bazalgette and Cresy divided the plans into seven distinct types and noted two main governing principles in the plans: first, 'the diversion from the Thames of the great mass of sewage now discharged into it'; and second, the combining of household sewage and rainfall in a single system of sewers.[41] They clearly favoured the adoption of an intercepting system and this reflects a shift in the overall goal of a new sewerage system: rather than emphasising the utilisation of London's sewage, Bazalgette and Cresy focused their attention on the purification of the Thames, which at that time still provided most of London's drinking water.[42] The scheme proposed by John M'Clean (a young engineer who later went on to become President of the Institution of Civil Engineers) was chosen as 'the best conceived and most practicable scheme' and, as an intercepting system, provides the basic elements of the plan eventually adopted and constructed. The report states that M'Clean's plan:

> ... is characterized by a well-devised system of intercepting sewers; in

> determining the situation and course of which a careful and elaborate study of levels has evidently been made. These intercepting sewers generally follow the direction of the main thoroughfares, and avoid any extensive interference with private property.[43]

The 'careful and elaborate study of levels' impressed Bazalgette and Cresy and their criticisms of other schemes tend to be couched in the same language: several of the plans are dismissed as 'vague speculative disquisitions' and all of the plans are criticised for providing estimated costs that were considered 'totally inadequate'.[44] Efficiency, practicability, thoroughness, and cost-effectiveness – these are the criteria by which the schemes are assessed. Bazalgette and Cresy also relegate the question of sewage utilisation, prevalent in many of the schemes, to a secondary concern: referring to M'Clean's plan, they state that:

> … though it leaves full scope for any operations designed for the preparation of sewage matter for agricultural purposes, that may be proved to be advisable, it does not render the efficiency of the entire scheme dependent upon the success of such operations.[45]

The question of sewage utilisation is sidelined precisely because it was one that could not be addressed in precise and practicable terms: it was an unresolved issue. The efficiency of the scheme becomes the primary objective and any proposed scheme is to be judged solely by this criteria.

When the 137 plans were considered for the second time on 15 March 1850, the commissioners were still unable to agree upon a suitable scheme to adopt and had already instructed Frank Forster, their consulting engineer, to prepare his own plan for the drainage of London.[46] Forster had worked initially as a mining engineer in south Wales; after 1830 he collaborated with Robert Stephenson on the design and building of the London to Birmingham railway.[47] He was appointed chief engineer to the Commission on 11 January 1850 and he submitted his plans for the southern drainage (Surrey and Kent districts) on 1 August 1850, and the northern drainage on 31 January 1851. These two reports formed the basis of Bazalgette's later system, which was largely an adapted and extended version of Forster's scheme.

In his report on the Surrey and Kent drainage, Forster starts by asserting the value of his scheme over those proposed in the competition. Referring to the competition plans he states:

> I have been … able to derive little or no practical assistance from any of them, which is to be accounted for, doubtless, by the necessarily defective data on which they were based …. I feel it is my duty to acknowledge … after consulting the block plans and subterranean surveys, a member of your Honourable commission, who kindly placed them in my hands during the preparation of the plan I have now the honour to submit.[48]

Forster's use of both the Ordnance Survey and subterranean survey plans (discussed in chapter 1) has enabled him to gain an intimate understanding of

London's topography and to position his scheme in relation to his predecessors: Forster implies that, while the competition plans were 'defective' in their method, his scheme is based on the scientific approach that characterised the production of the Ordnance Survey plans. This assertion was no doubt intended to invest his plan with an authority lacking in previous schemes.

Forster's plan divides the city into two distinct areas – north and south – that are defined by London's dominant topographical feature: the river Thames. Forster's reports on the northern and southern drainage are similarly presented: an illustrative map accompanies a long verbal description of each scheme. In his report on the northern drainage, Forster summarises the 'brief' given to him by the commission: the overall objectives of both schemes is:

> To keep the river Thames, within the Metropolitan districts, free from sewage at all times of the tide.
>
> To provide an escape by means of intercepting drains for so much of the sewage of the ... Thames as admits being so carried off.
>
> To raise all other sewage by artificial means, so as to deliver [it] into a main channel for removal.
>
> To construct the intercepting sewers as to secure the largest amount of effective drainage without having recourse to artificial means.[49]

These objectives, presented as specific problems to be solved, preclude the introduction of any ideal elements into the conception of the scheme. Consequently, Forster underplays the importance of sewage utilisation in his plan. In his report on the northern drainage, Forster had acknowledged that the undeveloped western areas of the city were 'eminently adapted for the application of sewage water as a manure', yet he also made clear that 'to guard against contingencies, I have provided for that area in calculating the sizes of the main low level line'.[50] In short, Forster only temporarily excludes this area of London from his scheme: the 'contingent' element – the uncertain value of sewage utilisation – prompts him to do this.

In his two reports, Forster also included coloured maps to explain his scheme in visual terms. **Pl. III** is a reproduction of the map that appeared in his report on the northern drainage. In the conception of his scheme, Forster divides the northern area of the city into three distinct regions: the area coloured red indicates where the intercepting sewers would drain by gravity alone, achieved by a natural slope in the land; and the area coloured light blue indicates where sewage and rainfall would require pumping by steam power. This light blue area was predominantly low-lying and therefore could not be drained by gravity alone; at a specified point along the intercepting sewer, the sewage would need to be pumped so that it could continue to the outfall draining by gravitation. The smaller dark blue areas indicate where only sewage would need to be pumped, the rainfall being allowed to drain into the rivers Lea and the Thames.

Forster's map plays a more central role in the presentation of his scheme than that used by Austin (**2.1**). Instead of using a skeletal plan with loosely defined areas, Forster's map ties in very closely with his verbal descriptions: with its precisely defined and distinct areas, this map represents a specific proposal for this site, as opposed to Austin's universal solution that could be applied anywhere. Forster clearly depicts the lines of his proposed intercepting sewers: the red lines indicating where sewage would be intercepted by gravity alone; the blue lines where the sewage would need to be pumped. Forster highlights the point at which the blue and red lines intersected as a convenient point to pump the sewage to the required height to drain by gravitation to the outfall, shown on the far right of the map. As described in his report, the red line – the main high-level sewer – would be 7 miles long and made up of four branches, shown on the map from right to left: the Hackney Brook branch; the Piccadilly branch; and the smaller Coppice Row and Regent's Park branches. The blue line – the low-level sewer – would be 8 miles long, its course staying near to the river and its two branches being, from right to left: the Isle of Dogs branch and the Bridge Street branch. From the junction of these two main lines, a large sewer, 4 miles long, would continue to the outfall situated at Barking Creek.

While Forster's map illustrates a scheme that is closely related to London's topography, it also represents that scheme encompassing and even bursting out of the city's boundaries: the red line of the outfall sewer, shown on the right of the map, extends well beyond the built-up area of London. Underpinning the vastness of this project is Forster's anticipation of a massive population increase, centred in the northern areas of the city.[51] Although Forster did not precisely define this expected increase he anticipates that the present water supplies in the city would soon double and consequently produce a corresponding rise in existing volumes of wastewater. Accordingly, he presents his scheme as both an accurate solution to existing problems and also encompassing and allowing for the potential growth of the city. Prediction – of urban growth, population increase, and the development of water supplies – is an important element within Forster's scheme. It is as much as a potential system, designed to accommodate future concerns, as it is a solution to existing problems.

Forster's map represents a clearly defined scheme, produced by a rational and scientific method of problem solving. Such a map fits into, what David Pike has referred to as, a 'genealogy of modernist space'.[52] In his analysis of Harry Beck's iconic 1931 map of the London Underground, Pike traces the origins of this type of 'abstract' mapping back to Napoleon III's plan of Paris – produced in 1853 as a blueprint for Baron Haussmann's redevelopment of the city – 'on which one saw traced in blue, in red, in yellow and in green … the different new [street] routes he proposed to take'.[53] If Pike sees one starting point of 'modernist space' represented in this map, Forster's earlier 1851 plan might equally be another. Referring to the Paris map and its proposed transformation of the city, Pike comments that 'such projects undertook, in the physical spaces of Paris, to control the chaotic, ungraspable reality of the modern city through color-coding, straight lines and diagonal cuts'.[54] Forster's scheme, represented in very much the same way as the Paris map, expresses this same goal for the underground sewer spaces of London. His scheme pictures rationally conceived lines of sewers, cutting their straight courses through the present tangle of underground sewer spaces, and providing the model for the future rationalisation of these spaces.

Forster's map marked the starting point of the development of a rationally conceived sewer system for London. During the period of the fourth sitting of the Metropolitan Commission of Sewers (formed in July 1852 after the third Commission collapsed due to financial difficulties), Forster resigned as chief engineer, his health affected 'by the harassing fatigues and anxieties of official duties'.[55] The commissioners were, yet again, divided over his plan and were forced to resign in October 1852, being superseded by the fifth Commission in November 1852. Bazalgette was appointed as Forster's successor on 26 November 1852.[56] Over the next six years, throughout the tenure of the fifth and the sixth Commission (formed after the fifth had yet again resigned office) and later the Metropolitan Board of Works, Bazalgette presented a series of reports outlining his gradual development of Forster's scheme. Bazalgette's early reports on both the northern and southern drainage specified the lines of intercepting sewers and began to present calculations as to their detailed elements. This increasing specification of the project culminated in two reports presented to the newly formed Board of Works in 1856. Bazalgette's plan for the southern drainage was read to the Board members on 3 April 1856 and the northern drainage on 22 May 1856.

Bazalgette's 1856 reports focused on the need to make an accurate prognosis of London's future growth. In his 1853 report on the northern drainage, Bazalgette made clear that any proposed scheme 'should be adapted not only to the present, but to the future wants of a rapidly increasing population'.[57] He predicted a 50 per cent increase in the current population, a far greater estimate than that made by Forster.[58] Bazalgette includes his predictions in tabulated appendices to his reports, with estimated population growth specified for individual districts within London.[59] He estimated a maximum population density, in the central areas of the city, of 36,000 persons per square mile, and he used this figure to estimate any future maximum density in these areas.[60] Thus, for the northern area alone, he estimated a population increase from 1,766,700 (in 1851) to a maximum of 2,513,900. Such prediction was, on the whole, uncertain: Bazalgette did not specify any timescale involved; rather, he based his figures on an assumed maximum density per square mile. However, by presenting his predictions in tabulated form (**2.3**), with figures given for every area through which his sewers would pass, Bazalgette gave the impression of accuracy and rigour in his calculations. For each area through which his sewers would pass, he estimated the future population increase and linked these with his other calculations, such as areas of each district, average rainfall, maximum flow of sewage, and the resulting sizes and inclinations of the sewers. These tables, providing detailed information for every area of the city, give a powerful impression of scientific rigour. If Forster's predictions appeared vague and generalised, Bazalgette's tables suggest an exact correspondence with his other calculations, even if their empirical basis was questionable.

Such prediction was also essential if constant enlargement of the system was

Northern Sewage

TABLE No. 2. MIDDLE LEVEL LINE

	POINTS OF INTERCEPTION ON MAIN LINE OF SEWER.	Estimated Area of Districts draining at Points of Junction.	Accumulated Areas of Districts.	POPULATION. Present Number.		POPULATION. Prospective Number, estimated at 30,000 per square mile.		SEWER On Prese
				On Separate Areas.	On Accumulated Areas.	On Separate Areas.	On Accumulated Areas.	On Separate Areas.
		Acres.	Acres.					Cubic Fee
1	At Stamford Brook (West Arm) at Acton ..	650	650	10,000	10,000	30,500	30,500	35
2	„ Old Oak Common-lane	350	1,000	200	10,200	16,400	46,900	1
3	„ Stamford Brook (East Arm)	450	1,450	6,000	16,200	21,100	68,000	21
4	„ Counters' Creek Sewer	450	1,900	10,000	26,200	21,100	89,100	35
5	„ Black Lion Toll Gate, Uxbridge-road ...	360	2,260	10,000	36,200	16,900	106,000	35
6	„ Ranelagh Sewer	1,850	4,110	30,000	66,200	86,700	192,700	104
7	„ King's Scholars' Pond Sewer	500	4,610	66,000	132,200	66,000	258,700	229
8	„ Regent-street Sewer	550	5,160	20,000	152,200	25,800	284,500	69
9	„ Gower-street Sewer	350	5,510	30,000	182,200	30,000	314,500	104
10	„ JUNCTION WITH PICCADILLY BRANCH ..	650	6,160	{ 40,000 } { 40,000 }	262,200	80,000	394,500	278
11	„ Fleet Sewer..........................	1,270	7,430	110,000	372,200	110,000	504,500	382
12	„ Goswell-street Sewer	350	7,780	70,000	442,200	70,000	574,500	243
13	„ City-road Sewer........................	500	8,280	30,000	472,200	30,000	604,500	104
14	„ Shoreditch Sewer	980	9,260	160,000	632,200	160,000	764,500	555
15	„ Globe-road	450	9,710	60,000	692,200	60,000	824,500	208
16	„ JUNCTION WITH ALDGATE BRANCH	450	10,160	{ 50,000 } { 30,000 }	772,200	80,000	904,500	278
	„ JUNCTION WITH NORTHERN HIGH LEVEL SEWER (or Hackney Brook), at the Dock Railway..............	100	10,260	10,000	782,200	10,000	914,500	35
	PICC							
	At Junction with Main Line	500	500	40,000	40,000	40,000	40,000	13
	AI							
	At Junction with Main Line	300	300	50,000	50,000	50,000	50,000	17

2.3 Extract from Joseph Bazalgette's tables of calculations included as an appendix to his report on the drainage of north London presented to the Metropolitan Board of Works on 22 May 1856.

to be avoided. Bazalgette envisages a system adapted to both present and future needs; the creation of a much larger system than was presently necessary would ensure its permanence in the urban fabric. Such a desire for permanence sheds light on Bazalgette's attitude towards the existing sewers of the city. In his 1856 report on the northern drainage, Bazalgette states that:

> It is better to construct the main lines first, and adapt the branches to them, rather than having first constructed the branches to be called upon to adapt the main lines to previously existing branches.[61]

Bazalgette views his intercepting sewers as massive and incontrovertible

SEWER AND BRANCHES.

at 5 Cubic Feet per Head [Per] Diem. On Prospective Population.		Maximum Flow of SEWAGE Running per Minute during Six Hours of the Day on the Prospective Population.		RAINFALL Running off per Minute, estimated at ¼ Inch in 24 Hours.		RAINFALL per Minute, and Maximum Sewage per Minute, added together.	
On Separate Areas.	On Accumulated Areas.	On Separate Areas.	On Accumulated Areas.	On Separate Areas.	On Accumulated Areas.	On Separate Areas.	On Accumulated Areas.
Cubic Feet.	Cubic Feet.	Cubic Feet.	Cubic Feet.	Cubic Feet.	Cubic Feet.	Cubic Feet.	Cubic Feet.
106	106	212	212	410	410	622	622
57	163	114	326	221	630	334	956
73	236	146	472	284	914	430	1,386
73	310	146	620	284	1,197	430	1,817
59	368	118	736	227	1,424	345	21,60
301	669	602	1,338	1,166	2,590	1,768	3,924
229	900	458	1,800	315	2,905	773	4,697
89	990	178	1,980	347	3,251	525	5,222
104	1,094	208	2,188	221	3,472	429	5,651
279	1,373	558	2,746	410	3,882	968	6,618
382	1,755	764	3,510	800	4,682	1,564	8,192
243	1,996	486	3,992	221	4,903	707	8,895
104	2,100	208	4,200	315	5,218	523	9,418
555	2,655	1,110	5,310	618	5,835	1,728	11,145
208	2,863	416	5,726	284	6,119	700	11,845
278	3,142	556	6,284	284	6,402	840	12,686
35	3,177	70	6,354	63	6,465	133	12,819
BRANCH.							
139	139	278	278	315	315	593	593
RANCH.							
174	174	348	348	189	189	537	537

Conditions of Proposed Sewers.		
Length.	Rate of Inclination.	Diameter.
Miles. Feet.	One in.	Feet. Inches.
1 , 2300	1,320	5 , 9
0 , 2050	1,320	6 , 6
1 , 870	1,320	7 , 3
1 , 1820	1,320	8 , 0
{ 0 , 4700 0 , 2250 }	1,320	{ 8 , 3 9 , 6
0 , 3900	1,320	5 , 6
0 , 2000	1,320	6 , 0
0 , 3230	1,532	6 , 0
0 , 3750	1,532	6 , 6
0 , 1730	1,532	7 , 0
{ 0 , 1150 0 , 880 }	2,640	{ 7 , 6 7 , 9
{ 0 , 730 0 , 820 }	2,640	8 , 0 8 , 6
{ 0 , 950 0 , 1550 }	2,640	8 , 9 9 , 0
{ 1 , 0 0 , 1020 }	2,640	9 , 3 9 , 6 }
0 , 1100	2,640	9 , 6
1 , 600	2,640	10 , 0
.... (See Table No. 1.)		
BRANCH. 2 , 0	1,408	3′ 9″ × 2′ 6″
RANCH. { 0 , 2250 1 , 2000 }	3520	{ 3′ 9″ × 2 6″ 4′ 0″ × 2′ 6″

interventions in the sewerage network of the city that would act as catalysts for the future improvement of that network. As such, his intercepting sewers are particular 'tools' placed in the underground urban fabric designed to eventually transform it. Such implied perfection contrasts sharply with Chadwick's vision, which required a wholesale replacement of the old with the new. Consequently, for Bazalgette, the efficiency of his new lines of sewers is not dependent on the removal of the present sewers. Indeed, after the construction of the intercepting sewers in the 1860s, a scrupulous monitoring of all proposals for new local sewers guaranteed that they were compatible with the new system already potentially in place.[62] Such foresight ensured an ever-evolving system and an ever-tightening network, rigorously monitored by those who had inaugurated that network in a massive and irreversible intervention. The intercepting system of sewers thus set the circulation of waste in the city firmly in the control of the architects of that system, who strictly controlled its gradual development after the imposition of the main drainage system.

Bazalgette begins his 1856 report on the northern drainage by explaining the general principles of drainage itself: normally land would drain naturally into nearby watercourses, but London is:

> A city of enormous magnitude and population … intersected by railways, roads and canals, overcrowded with traffic, and consisting in great measure of heavy building.[63]

Bazalgette stresses that in this unnatural and chaotic environment a system of complete drainage was needed with intercepting sewers to 'provide an artificial and elevated outfall, and to create new valley lines across the ridges which separate the natural valleys'.[64] This conception is governed by the goal of keeping sewage out of the Thames in the metropolitan area. To achieve this, Bazalgette largely adopts Forster's divisions of drainage for both the northern and southern areas. Referring to two maps attached to his 1856 reports, one for the north side (**Pl. IV: a**) and one for the south (**Pl. IV: b**), Bazalgette explains his drainage divisions: he divides the southern area of London into two areas: the high-level area (19.75 square miles), coloured pink, to be drained by gravitation; and the low level area (22 square miles), coloured blue, to be drained by pumping. Two main intercepting sewers – the low and high level sewers – would drain both of these areas; the pumping station being located at Deptford Creek – the point where both sewers joined. The southern outfall sewer, running from Deptford Creek to the outfall at Erith, would need to be pumped again at the outfall, due to the low-lying land through which it would pass. Bazalgette's plan for the southern drainage is essentially a much-enlarged version of Forster's scheme, which took into account the rapid development of south London in the 1850s.

Turning now to the northern drainage, Bazalgette specifies four drainage divisions: the high level area (9.68 square miles), coloured yellow; the middle level area (17.64 square miles), coloured pink; the low level area (10.8 square miles), coloured blue; and a new western division (21.45 square miles), coloured brown and green. Bazalgette defines these areas more precisely than Forster and clearly presents their respective sizes on his maps. The courses of the intercepting sewers are similar to Forster's, but with extensions in the western districts, comprising the Acton line, and the Brentford line and its Fulham branch. Bazalgette's reports describe in detail the courses of these lines of sewers, charting their trajectory street-by-street from source to outfall, while the accompanying maps presents a clear and simplified picture of this detail. For both maps, Bazalgette employs a different base plan from that used by Forster. His system is superimposed on top of the index plan of the Ordnance Survey made for the Metropolitan Commission of Sewers in 1850 (**1.1**). Characterised by a lack of topographical detail, the Ordnance Survey index plan instead highlights the proposed lines of sewers and the areas through which they would pass. The thickness of the sewer lines, together with their clear labelling, is bolder and clearer than in Forster's map. Only topographic landmarks, such as main thoroughfares and parks, serve as guidelines as to their proposed courses. Similar prominence is also given to other similar 'lines' within the city – the railways. The superimposition of Bazalgette's scheme onto the Ordnance Survey index plan serves to link both the northern and southern schemes: it presents a unified representation of London, epitomised by its grid-like rectangles seamlessly linked together (see chapter 1). The scientific rationale behind this plan is here directly equated with Bazalgette's scheme, superimposed effortlessly onto it. Consequently, his scheme might be viewed as a realisation of the potential of the Ordnance Survey map, as originally and with hindsight, ironically, stated by Chadwick in 1848.[65] Bazalgette's sewers, presented as lines cutting through the city, are welded to the overall grid of the index plan with its scientific basis. As such this map contrasts with Forster's, where the sewers are more closely associated with streets and buildings in the city. In short, here is a vision of rationality cutting through and reorganising the chaotic underground world of the city.

Rational Flows and Bodies

These visual representations are complemented, in Bazalgette's reports, by a concentration on the detailed elements of his system and the ways in which it would perpetuate a constant flow of wastes out of the city. This concern with a new kind of 'circulation', more properly defined as flow, is increasingly emphasised by Bazalgette in the 1850s. In his early reports, on both the northern and southern drainage, Bazalgette begins to define the proposed sizes and inclinations of the sewers, as well as the velocity of current required in those sewers to create a constant flow of sewage 'clear from deposit'.[66] He defines the minimum velocity of current within any part of the intercepting sewers as two-and-a-half miles per hour,[67] intended to be uniform throughout the sewers and 'sufficient to make them self-cleansing'.[68] If Chadwick had envisaged a self-cleansing sewer achieved through the rigid control of external variables, Bazalgette concentrates on the design of the sewers themselves. His 1854 report on the northern drainage contained tables of statistics outlining the proposed sizes, lengths and inclinations of his intercepting sewers.[69] Bazalgette's statistics were given credence by the testimony of two eminent consulting engineers, Robert Stephenson and Sir William Cubitt, who praised them as 'clearly and minutely set forth'[70] presenting a 'simple and efficient system of drainage'.[71]

Bazalgette's conception of flow is very different from Chadwick and Austin's notion of circulation. Rather than establishing a grand circular movement – from households to fields and back again – Bazalgette's intercepting system intended to transport sewage from one point to another in a linear flow, eventually flushing it into the river as a waste product. For Chadwick, this element of waste was not only potentially economically disastrous but also unnatural and artificial, going against the divinely instituted laws of nature. Therefore, in Chadwick's eyes, Bazalgette's eschewing of sewage recycling was not only unsound economically but also morally: Bazalgette had failed to obey God's self-evident laws. Bazalgette's conception also went against Chadwick and Austin's focus on the regulation of the domestic sphere and the human body. If Chadwick described his system as 'arterial-venous' – that is, directly modelled on the workings of the human body – Bazalgette makes no such analogies in his descriptions of the main drainage system; Chadwick's organic circular movement, modelled on the body's veins and arteries, is superseded by a rational series of flows in abstractly conceived subterranean spaces. In addition, and in direct contrast to Chadwick and Austin, Bazalgette makes no mention in his reports of domestic sanitation and bodily practices: in effect, the body is entirely excluded from his conception. This policy of non-intervention in the private sphere was no doubt a consequence of simple practicalities; but it also represents a fundamental shift in the understanding of sewer spaces in the city. In Bazalgette's conception, the body 'disappears' and is instead replaced by a series of fragmentary yet homogenous spaces, defined by the tenets of science and efficiency and increasingly detached from the body. In effect, a new 'body' is created: a body of abstract knowledge that now defines the way in which sewers and their flows would be understood.

The extent to which Bazalgette's rational conception of London's sewers had superseded Chadwick and Austin's organic idealism is demonstrated in the handling of opposition that arose to his intercepting scheme. Despite the endorsement of Cubitt and Stephenson, Bazalgette's scheme was challenged by other proposals sent to the Metropolitan Commission of Sewers that attempted to re-establish a form of Chadwickian idealism. In October 1854, the engineer John Roe, who had been closely associated with Chadwick, outlined his own plan for the drainage of London, focusing on the utilisation of the city's sewage.[72] Roe's intercepting scheme proposed different lines of sewers to those outlined by Bazalgette, and he also included his own detailed calculations that specifically challenged Bazalgette's figures. Despite a detailed rebuttal of Roe's plan by Cubitt, Stephenson, and Bazalgette himself,[73] Roe persisted and put forward a further report on 15 January 1855 accusing Bazalgette of making 'blunders' and insisting that his own scheme contained 'no errors in [its] calculations'.[74] Bazalgette's subsequent report, presented to the Commission on 22 January 1855, was clearly meant as a definitive response to Roe's challenge.[75]

In his report, Bazalgette laid out, in minute detail, the quantities and costs of materials required for his system; the flow, inclination, depth and sizes of all the sewers; the hydraulic pressure in the sewers; the pumping power necessary; and details of all other costs, including labour and supervision fees. Such an enormous range of detail was clearly intended to validate Bazalgette's working method and to silence opposition once and for all. In terms of its content, this report represented the antithesis of Chadwick and Austin's universal system that was guided by a theological-based cosmic vision: instead, Bazalgette eschews all speculation, whether theological or otherwise, and instead presents a minutely specified series of interconnected spaces, all scientific in their conception and designed to facilitate a perpetual flow in the whole system. The very fact that the question of scientific method had now moved centre stage, whether that applied by Bazalgette or his Chadwick-influenced opponents, points to the level of authority invested in the very notions of precision and rationality, as opposed to any wider economic, theological or bodily concerns. In the end, Bazalgette's conception won the day and, in the next two chapters, I explore the translation of this abstract conception into a concrete reality: just how were these vast rational tools to be constructed in the chaos of the city and what would its citizens make of them?

Section II: Construction

3

Contracts and Construction

To transfer a substantial engineering project to a built reality requires hundreds of technical representations – sketches, orthographic plans, schematics, diagrams, perspective drawings and models, to name only the most common. In their overtly technical language, these representations are usually assumed to be directed only at audiences who can interpret them, namely engineers, architects and contractors.[1] Yet, in any engineering project, especially in the field of civil engineering, technical representations are tools of communication directed at a wide variety of audiences other than fellow-engineers and contractors.[2] Civil engineers must devise means of imposing their design over a natural environment that can only be known incompletely – before, during and after the project. To successfully achieve this, they must form alliances with many different parties: from those who authorise and finance the project; the contracting firm that will execute their plans; the various parties who have vested interests in the land impinged upon by the project; to the wider public who are supposed to benefit from it and in many cases, indirectly or otherwise, pay for it. In short, civil engineers, far more than their mechanical counterparts, must, through their representations, gain the confidence and cooperation of a host of diverse audiences if they are to have any chance of successfully transforming their designs into a built reality.

This chapter considers the role of technical representations in the building of the main drainage system in the early 1860s. I use the term technical representations to mean the written and pictorial documents used by Bazalgette to transfer his design into a built reality. My primary focus will therefore be on the contract – composed of engineering drawings and an accompanying specification – which was the key document that mediated the relationship between engineer and his most important ally, the contractor. I will also pay close attention to the variety of audiences beyond the contactor, to which these documents were directed, including: the organisation to which Bazalgette was responsible, the Metropolitan Board of Works; those who funded the project; those parties directly affected by construction; and the wider public, who would inevitably be responsible for paying for the project. The result will be to construct a fuller picture of the social context in which the main drainage project was constructed and the crucial role played by the contract in mediating social relations of many kinds, a perspective that is absent in the existing literature on the subject.[3]

The Sanction of Bazalgette's Plan

Chapter 2 focused on the development of Bazalgette's main drainage plan up to 1856, when it was presented and approved by the Metropolitan Board of Works. However, Bazalgette's success was short-lived. Although the 1855 Metropolitan Local Management Act, which created the Board of Works, charged it with the task of constructing a citywide sewerage system, the Act also specified that any projects costing more than £50,000 would need to be approved by the Crown-appointed First Commissioner of Works, a post held in 1856 by Sir Benjamin Hall (1802-67).[4] On 3 July 1856, Hall rejected Bazalgette's plan stating that, in his opinion, the positions of the outfalls were not sufficiently far enough away from the metropolitan area to prevent sewage from returning on the incoming tide.[5] There followed a protracted and often bitter dispute between Bazalgette and Hall which resulted in further plans being drawn up by both Bazalgette and independent referees appointed by Hall.[6] By the time Bazalgette presented his final plan on 6 April 1858, with the assistance of the engineers George Parker Bidder and Thomas Hawksley, the question was still unresolved.[7]

The dispute was brought to a rapid conclusion during the summer of 1858 – the hottest on record – when the notorious and much-documented 'Great Stink' enveloped London.[8] The swift passing of the Metropolitan Local Management Amendment Act of 2 August 1858 – just two weeks and four days after its first reading in Parliament – was a result of both the fear of the smell emanating from the Thames and also a more accommodating First Commissioner of Works, Lord John Manners. The new Act gave the Board of Works authority to build whatever scheme it approved and allowed it to borrow three million pounds from the Bank of England, which would be paid back over 40 years by means of a three pence rate imposed on all properties within the metropolitan district.[9] In addition, the Act removed the power of the First Commissioner to veto the scheme. From now on Bazalgette's drawings would no longer be subject to the scrutiny of the Government; in effect, the most powerful agent in Bazalgette's former hierarchy of responsibility was removed, allowing him a great deal of autonomy in the subsequent production and tendering of contracts.

Tendering

When, in 1888, Bazalgette explained the process by which the main drainage contracts were tendered, he was giving evidence to a Royal Commission set up to investigate the working practices of the Metropolitan Board of Works.[10] Despite his insistence that the tendering process was impartial and governed by meticulous procedures, in 1888 Bazalgette faced accusations of favouritism in the awarding of contracts. On one level the accusation seems accurate. Of the 27 main drainage contracts drawn up from 1859 to 1865, the contractor

George Furness (1820-1900) was awarded two and William Webster five; Bazalgette favoured both firms for their reliability and reused them many times in the course of his career, later awarding them both prestigious contracts for the Thames Embankment.[11] Not only did Bazalgette favour well-known contractors over those that offered the lowest tender, but, in the case of the Abbey Mills pumping station, he awarded the project to Webster without having drawn up a contract at all, bypassing the apparently uniform process he described in such detail in 1888 (see chapter 5).[12]

In the case of the contract for the northern outfall sewer, the largest and most important of all the main drainage contracts, the meticulous procedures described by Bazalgette in 1888 were apparently followed, however. Once the deadline for tenders had expired on 18 October 1860, the tenders were deposited in a box on the afternoon before the Board meeting on the following day, where they were opened by the clerk and handed to its Chairman, who formally announced them to the Board members.[13]

The northern outfall sewer contract, made up of 51 drawings and an 85-page specification, was drawn up from 1859 to 1860 after eleven of the other contracts had already been successfully tendered.[14] The production of contract drawings – over 400 for the 27 contracts – in Bazalgette's office took place within a strict hierarchy of labour with draughtsmen mainly fulfilling the role of copyists. They were paid on a weekly basis and, due to the pressure of tight deadlines for contracts, were often required to work unrestricted hours set by Bazalgette.[15] They were responsible for producing contract drawings under Bazalgette's supervision and for overseeing their reproduction, in this case by lithography. All of the original main drainage contract drawings, now held at the London Metropolitan Archives, were meticulously coloured by Bazalgette's team of draughtsmen, whereas all reproductions were monochrome. Significantly, it was the reproductive copies that formed the basis for the contract that was advertised to potential tenders as well as for reproductions in periodicals and newspapers. This means that the original drawings must have been coloured *after* reproduction: that is, colour is not a necessary aspect of their contractual function. Rather, the meticulous colouring of all of the original contract drawings reflects the importance Bazalgette invested in them as aesthetic objects, the result of which would be an enhancing of his own professional status and reputation among his peers.

After Bazalgette and his draughtsmen had completed the drawings, a specification was drawn up detailing all aspects of construction, materials and costs. The whole was then presented to the members of the Board of Works for approval before being advertised to potential tenders. In the case of the northern outfall sewer, the drawings were presented to the Board on 10 August 1860, attended by 27 of its members.[16] Both lithographed copies and specification were still being prepared and Bazalgette hoped to submit them to the Board within a week. In fact, the Board members approved the drawings and authorised Bazalgette to issue advertisements for the tenders immediately, the deadline for applications expiring on 4 October. Bazalgette's audience at these meetings were representatives from London's vestries and district boards, who would have had very limited knowledge of how to interpret engineering drawings. The fact that Bazalgette presented colour drawings to the Board suggests that aesthetic appeal was an important factor in impressing and convincing his audience, especially as the use of colour was unrelated to the contractors' needs. Whatever the reason for this use of colour, the result was that the viability of Bazalgette's drawings in realising the project was invariably left unquestioned by the Board members.

Once this audience had been negotiated, Bazalgette addressed his key potential allies in the realisation of his project: the contractors. For all of the 27 main drainage contracts, Bazalgette employed a system known as 'measure and value', in which every detail of the project was specified in advance, the contractor being repaid very slowly as construction progressed. In the early-19th century this type of contract gradually superseded the traditional 'after measurement' contract, where the specification would be drawn up after construction (and from which the cost of the project would be ascertained).[17] The expansion in the scale of engineering projects in the early-19th century stimulated a corresponding intensification of competition amongst contractors. By mid-century a small number of very large contracting firms had succeeded, by the process of competitive tendering for large-scale railway contracts, in dominating the building industry.[18] After William Rowe, the original contractor for the northern middle-level sewer, went bankrupt soon after gaining the contract in February 1860, Bazalgette began to favour larger contracting firms, who did not always offer the lowest tender.[19] In the case of the northern outfall sewer, he even extended the deadline for tender, at the request of Thomas Brassey, the owner of one of the largest contracting firms of the 19th century.[20] Indeed, prospective contractors were strongly advised to purchase copies of the drawings and specification for the considerable fee of five pounds that, according to Bazalgette, was 'to secure that none but *bona fide* tenderers shall send in their tenders'.[21]

On 19 October 1860, the tenders for the northern outfall sewer were opened and examined. Ten contracting firms put in bids ranging from £625,000 from George Furness to £699,500 from Mr G. Todd.[22] Out of these, four were already employed on other main drainage contracts: Thomas Brassey, William Webster, William Dethick, and William Moxon.[23] All of these bids were compared with Bazalgette's estimate of £635,000, calculated with the aid of the Board's quantity surveyor and solicitor. Like Brassey and Webster, Furness had already gained extensive experience in railway contracting, working in the 1840s and 1850s on projects in England, France and Brazil.[24] One factor that may have influenced the Board's decision to accept Furness's tender was his proven ability to carry out several contracts simultaneously, a practice common in railway construction and one adopted by Bazalgette for the main drainage system. However, it is clear that Bazalgette's inclination to favour reliable contractors was tempered by the Board's concern with expense; and yet, like the meeting in which Bazalgette presented the contract, no discussion is recorded between the Board members, indicating that their confidence in Bazalgette's judgement was remarkably high. Like Webster and Brassey, Furness was to prove an important ally in the successful realisation of two other main drainage contracts: the northern outfall reservoir and part of the Thames Embankment. Before the contract was signed and the 'common seal of the Board' affixed to it, two reliable sureties confirmed the contractor's financial status.[25] Furness also signed each of the 51 contract drawings and the specification to confirm his contractual obligations and responsibility for the entire project.

Drawings

Both Furness and the other nine bidders for the northern outfall sewer contract would have relied upon the 51 contract drawings, in conjunction with the specification, to determine their respective tenders. Although Furness was not an engineer, he was elected an Associate of the Institution of Civil Engineers in 1864 – recognition of his extensive experience and knowledge of engineering practice.[26] His early experience in railway contracting prepared him well for interpreting Bazalgette's drawings. Both the index map (**3.1**) and index section (**3.2**) were drawings common to all large-scale civil engineering projects, which, due to their enormous geographical extent, required concise overviews. The index map shows the proposed course of the sewer, in this case a distance of five miles and 1,400 feet. In the index section the otherwise imperceptible inclination of the sewer (two-feet per mile) is shown by employing vastly different horizontal and vertical scales (in **3.2** the horizontal scale is 880-feet to one-inch; the vertical scale is eleven-feet to one-inch). This disparity of scale serves to distort the topographical features, exaggerating vertical variations – seen as prominent peaks and troughs along the route of the sewer. Both the index map and section, as well as providing a complete overview of the project, also indicate to the contractor the type of obstacles that would be encountered during construction: in this case roads, rivers and railway lines.

The information presented in the index map and section was greatly expanded in the general map and section: drawings also common to all of the main drainage contracts. At ten times the scale of the indexes, these drawings rendered topographic details such as buildings, rivers and vegetation, ground and water levels, information concerning land ownership, and references to other drawings in the contract. An extract of the northern outfall sewer general plan shown in **Pl. V** formed a small part of the original coloured sheet that measured approximately ten metres in length; the subsequent lithographic copies of these drawings, intended for the contractor's use, were divided into many separate sheets. The original coloured sheet was almost certainly one of the drawings presented to Board members to approve the contract in August 1860. The use of colour in both the general plan and section is more elaborate than in the original index drawings, with topographic details such as buildings, rivers and vegetation immaculately rendered in browns, blues and greens respectively, with the intended line of the sewer shown in red. Additional detail not shown on the index drawings includes information, prominently labelled, concerning land ownership (most of the land shown is labelled 'East London Waterworks Company'). For Board members, each of whom represented the

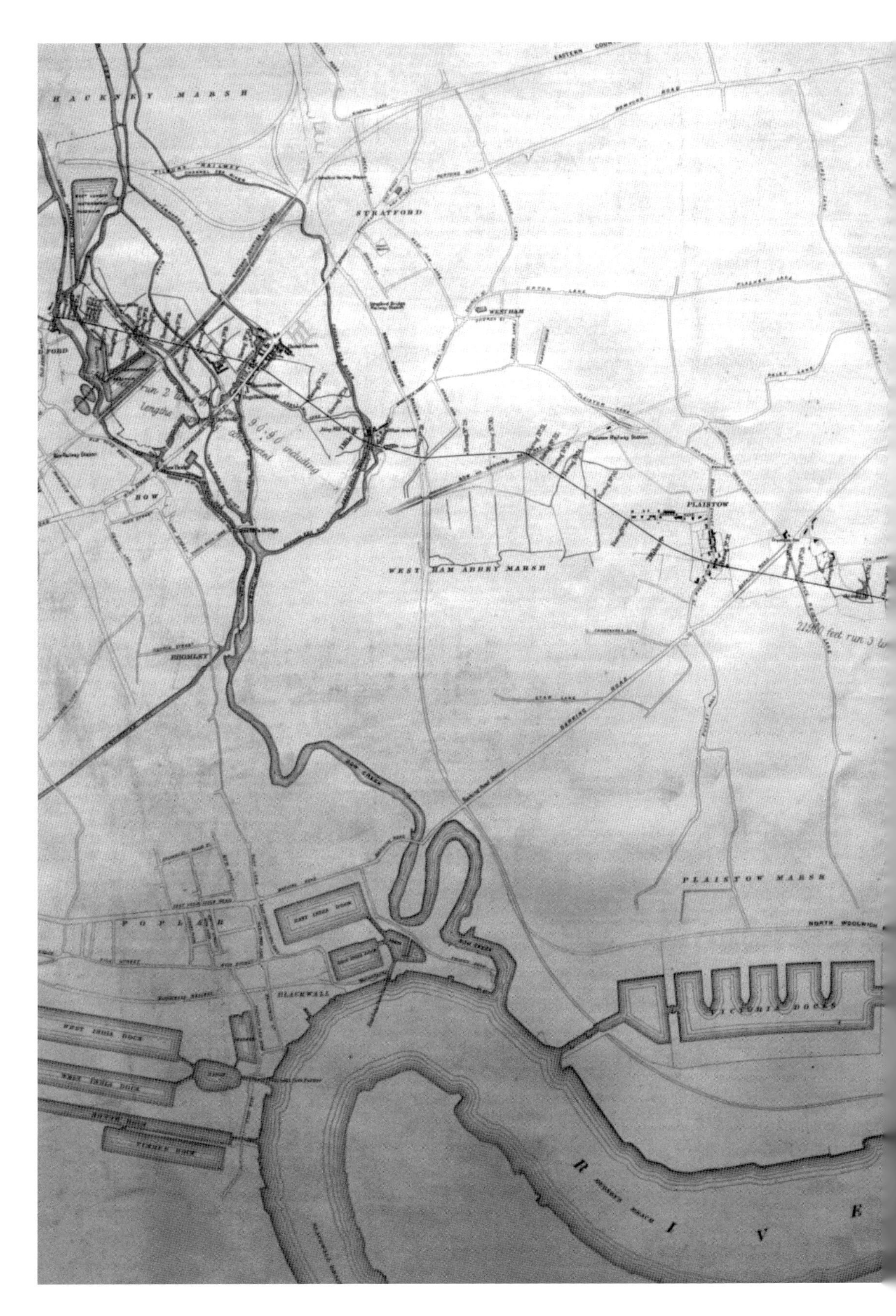

3.1 The first drawing in the Northern outfall sewer contract showing the proposed course of the sewer across the Essex marshes.

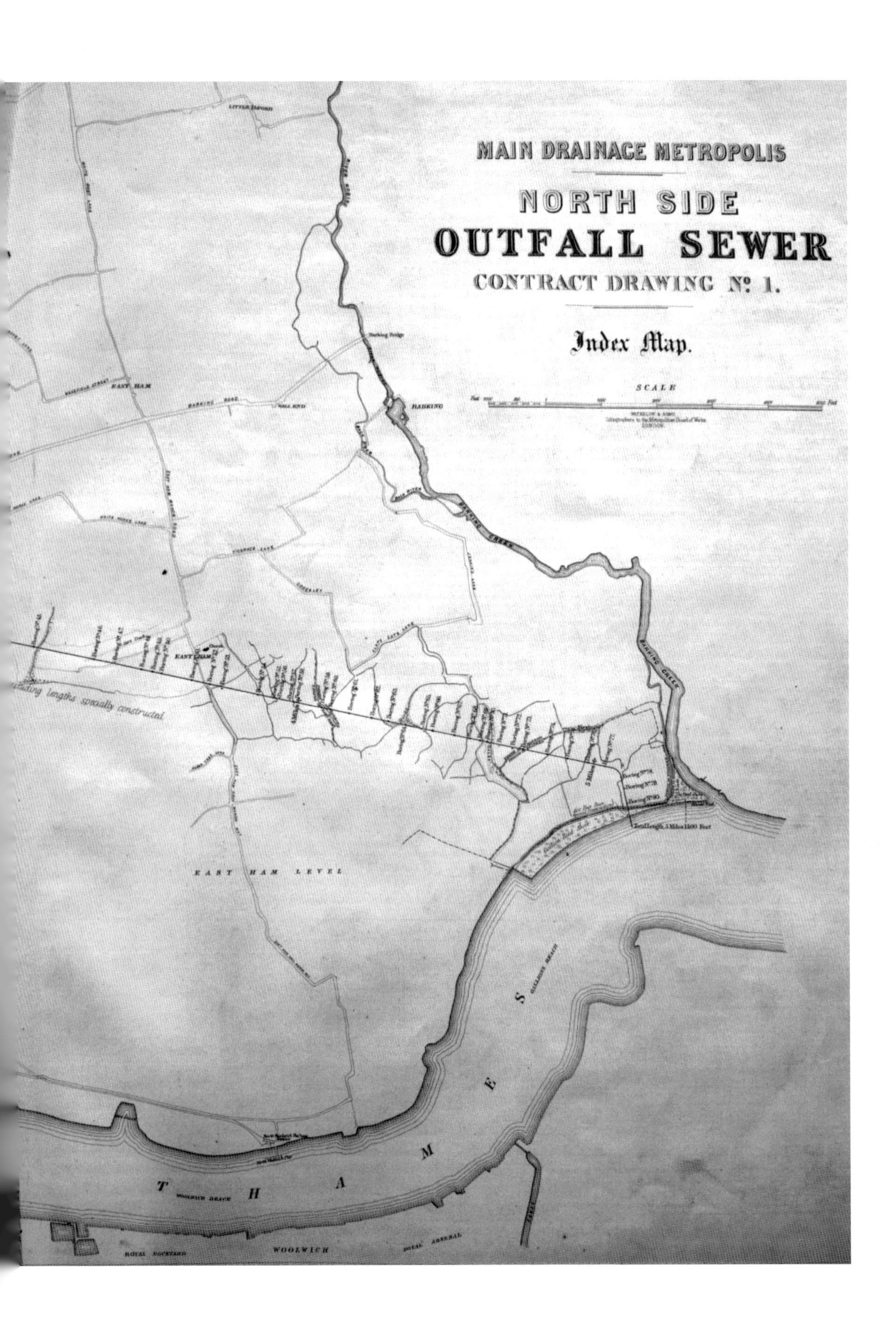
MAIN DRAINAGE METROPOLIS
NORTH SIDE
OUTFALL SEWER
CONTRACT DRAWING No. 1.
Index Map.
SCALE
BARKING
EAST HAM
BARKING ROAD
BARKING CREEK
EAST HAM LEVEL
T H A M E S
WOOLWICH REACH
ROYAL DOCKYARD
WOOLWICH
ROYAL ARSENAL

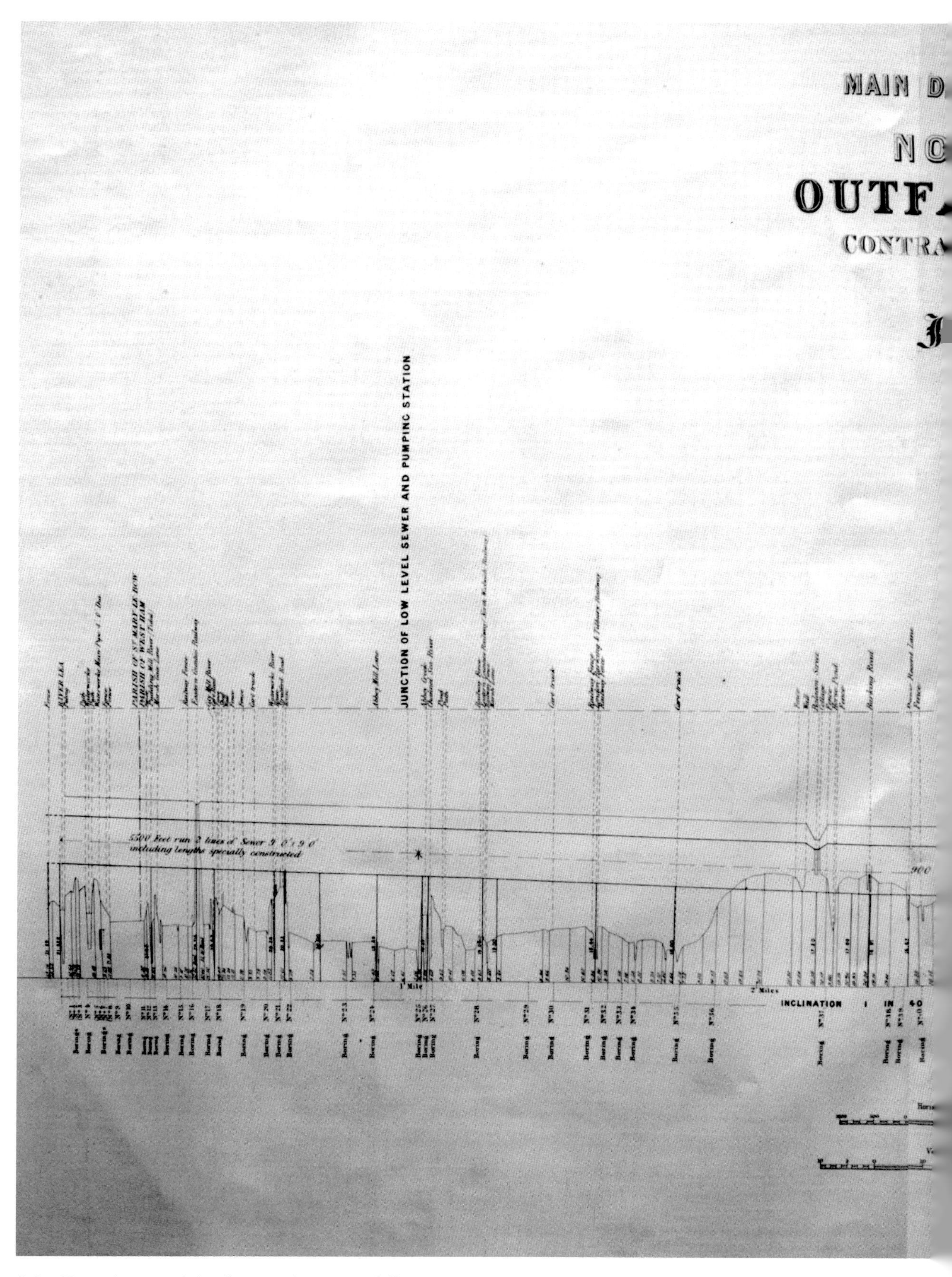

3.2 Drawing no. 2 in the northern outfall sewer contract, a sectional view of the entire length of the sewer showing the vastly different horizontal and vertical scales (the horizontal scale is 880-feet to one-inch; the vertical, eleven-feet to one-inch).

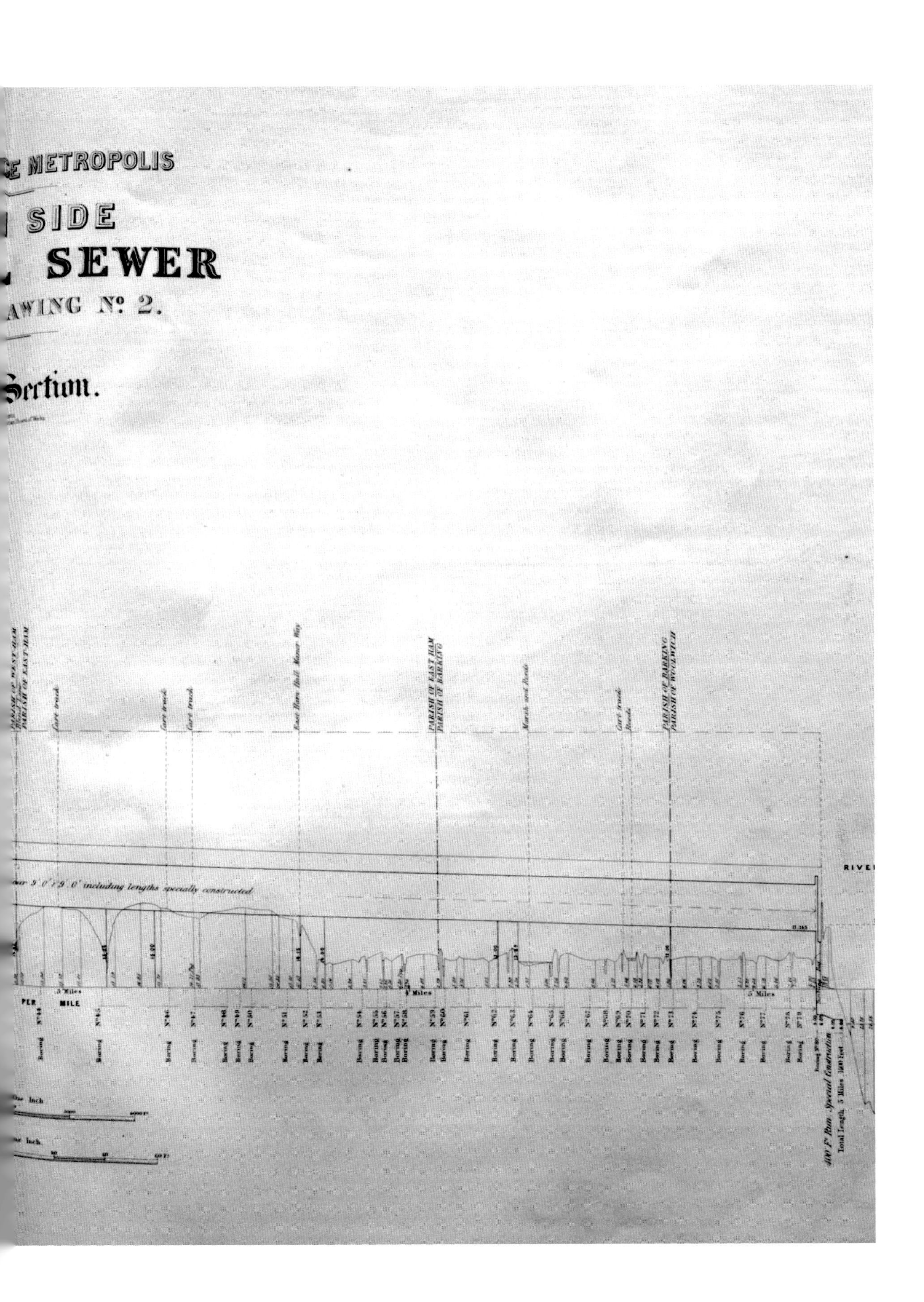

METROPOLIS
SIDE
SEWER
AWING No. 2.
Section.
PARISH OF WEST-HAM
PARISH OF EAST-HAM
Cart track
Cart track
Cart track
East Barn Hall Manor Way
PARISH OF EAST HAM
PARISH OF BARKING
Marsh and Roads
Cart track
Roads
PARISH OF BARKING
PARISH OF WOOLWICH
RIVE
3 Miles
4 Miles
5 Miles
PER MILE
400 F. Run, Special Construction
Total Length, 5 Miles 1240 Feet

interests of individual vestries and district boards in London, such information would have been of the utmost importance: it highlighted which landowners would be affected within their respective areas of jurisdiction. For the contractor the situation was different; his primary concern lay not with the interests of landowners, but with production; it is unlikely that either index or general plans and sections would have provided him with a great deal of useful information, particularly as the course of the sewer had already been marked out with iron fencing before the contract had even been advertised.[27]

It is in the detailed drawings of specific parts of the project – making up the bulk of the 51 contract drawings – that the needs of the contractor were most directly addressed. Unlike the index and general maps and sections, these drawings make use of three-view orthographic projection – that is, plan, elevation and sectional views – in the tradition of mechanical engineering drafting from the late-18th century onwards.[28] This technique enables various views of individual parts of the project to be arranged on single sheets, explicitly addressing the party who would have the knowledge to interpret and execute them: namely, the contractor. Drawings 5 to 12 detail the method of construction for the proposed viaduct over the River Lee that would carry the twin lines of the northern outfall sewer. **3.3** shows drawing 5, depicting the general arrangement of the intended works at the River Lee. The plan view of the viaduct, seen at the top-left of the drawing, represents the 'key' to the other drawings scattered over the rest of the sheet – sectional views that clarify 'configurations that can otherwise be deduced only from hidden lines'.[29] The River Lee was only one of many obstacles encountered in the planning of the northern outfall sewer: other drawings picture more aqueducts over the Waterworks River, the Pudding-Mill River, Abbey Creek, the Channelsea and City Mill Rivers, and numerous minor watercourses; and bridges and tunnels over many roads and railways, the most complex being the raising of the Stratford Road 10-feet above the sewer embankment.

As outlined above, during the tendering process, the audience for the main drainage contracts comprised both those directly associated with the production of the contract – Bazalgette, his draftsmen, quantity surveyors and solicitors – and also those who were part of the decision-making process – the members of the Metropolitan Board of Works, the group of bidding contractors, and finally the successful bidder. However, the contract drawings were also subject to the scrutiny of more disparate audiences not directly involved in either the drafting or decision-making processes. If the 1858 Metropolitan Local Management Amendment Act had limited the power of Government to interfere with the realisation of the main drainage system it had not removed it altogether; for each of the 27 main drainage contracts, permission had to be sought and granted by the Secretary of State for the Board to compulsorily requisition property and lands. In all cases this permission was granted, but was often strongly contested by more important landowners, resulting in delays in the commencing of construction.[30] In the case of the northern outfall sewer, the contract drawings, particularly the general plan, outlined in detail the various occupiers of land over which the sewer would pass. Throughout the period of construction, negotiations for land purchase and compensation occupied much of the time of the Metropolitan Board of Works – more especially, their hard-pressed solicitor, who dealt with all negotiations and settlements.[31] In the case of the northern outfall sewer, 30-feet of land either side of the sewer also had to be purchased to allow for access during construction; in December 1860, the unlucky residents and owners were informed and the solicitor drew up the necessary agreements.[32] These land negotiations reveal the Board of Works asserting their authority often in the face of understandable objections from property owners; however, the straight line of the sewer was no respecter of persons, and the courts consistently upheld its authority as a 'public' project.[33]

Unlike their counterparts in mechanical engineering, civil engineers were often required to address this nebulously defined 'public'. The main drainage system was, from the start, explicitly deemed a 'public' project, in that it claimed to benefit, through improved sanitation, a specific group of people: in this case Londoners, who would also eventually pay for the project in the form of property rates. It is not surprising, then, that throughout the planning and construction of the main drainage system, contract drawings were important mediating representations between Bazalgette and the Metropolitan Board of Works and the wider metropolitan population. On the one hand, Bazalgette used contract drawings to illustrate the many talks he gave on the main drainage system to specialist audiences of engineers; on the other he exposed them to a much wider public audience, by exhibiting some unspecified contract drawings in the 1862 International Exhibition in South Kensington, and providing others for London's periodical press to reproduce.[34] When the *Engineer* published a series of drawings and specification extracts from the 1865 contract for the Abbey Mills pumping station, it proudly proclaimed that 'so complete a description of any engineering work has never before appeared in the pages of a weekly journal'.[35] Not only was the journal promoting the main drainage system – and Bazalgette's prowess – to a specialised audience that had the skills to read these drawings; it was also celebrating its own representation of the project.

During the 1860s, the *Illustrated London News* also published many wood-engraved topographical views of the construction of the main drainage system. Most of the newspapers' illustrations were dramatic representations of the building works, which I examine in more detail in chapter 4, but some were also direct reproductions of Bazalgette's contract drawings.[36] When the newspaper represented the construction of the northern outfall sewer in 1864, it uniquely combined these within the same engraving (**3.4**).[37] In the full-page engraving, the large topographical image shows the outfall of the sewer – the hovering viewpoint serving to reveal the vast scale of construction seen in the mid- and far-distance, made up of a confusion of rubble, machinery and workmen. Below this dramatic image is a sectional view of the works based on part of drawing 43 from the northern outfall sewer contract. However, the context and content of this reproduction are very different from its source image: the engraving greatly simplifies the contract drawing by removing all measurements and much of the topographical labelling. In addition, the engraved image is divorced from its

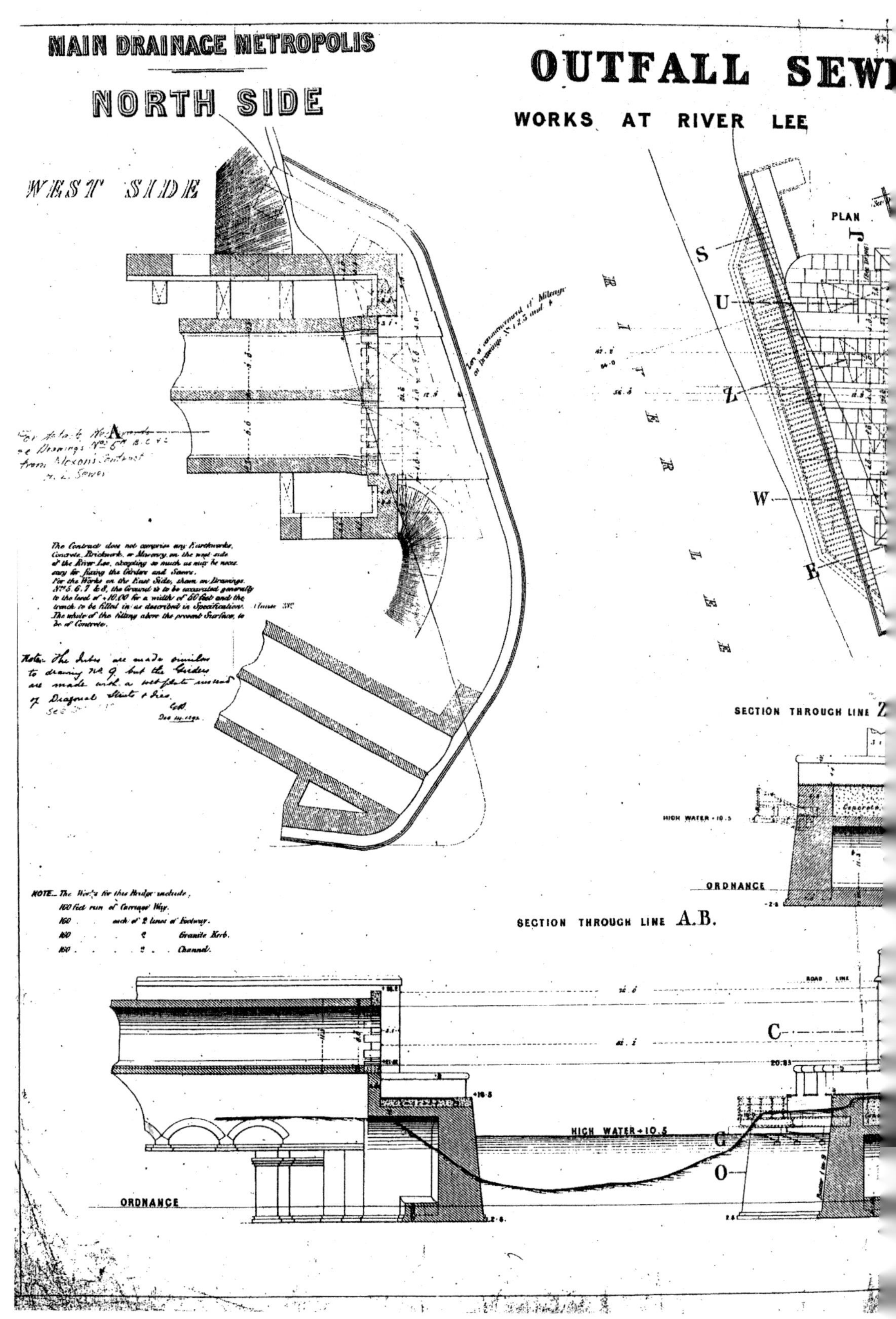

3.3 Drawing no. 5 in the northern outfall sewer contract showing the intended works at the River Lee, one of the major obstacles to be negotiated in the building of the sewer.

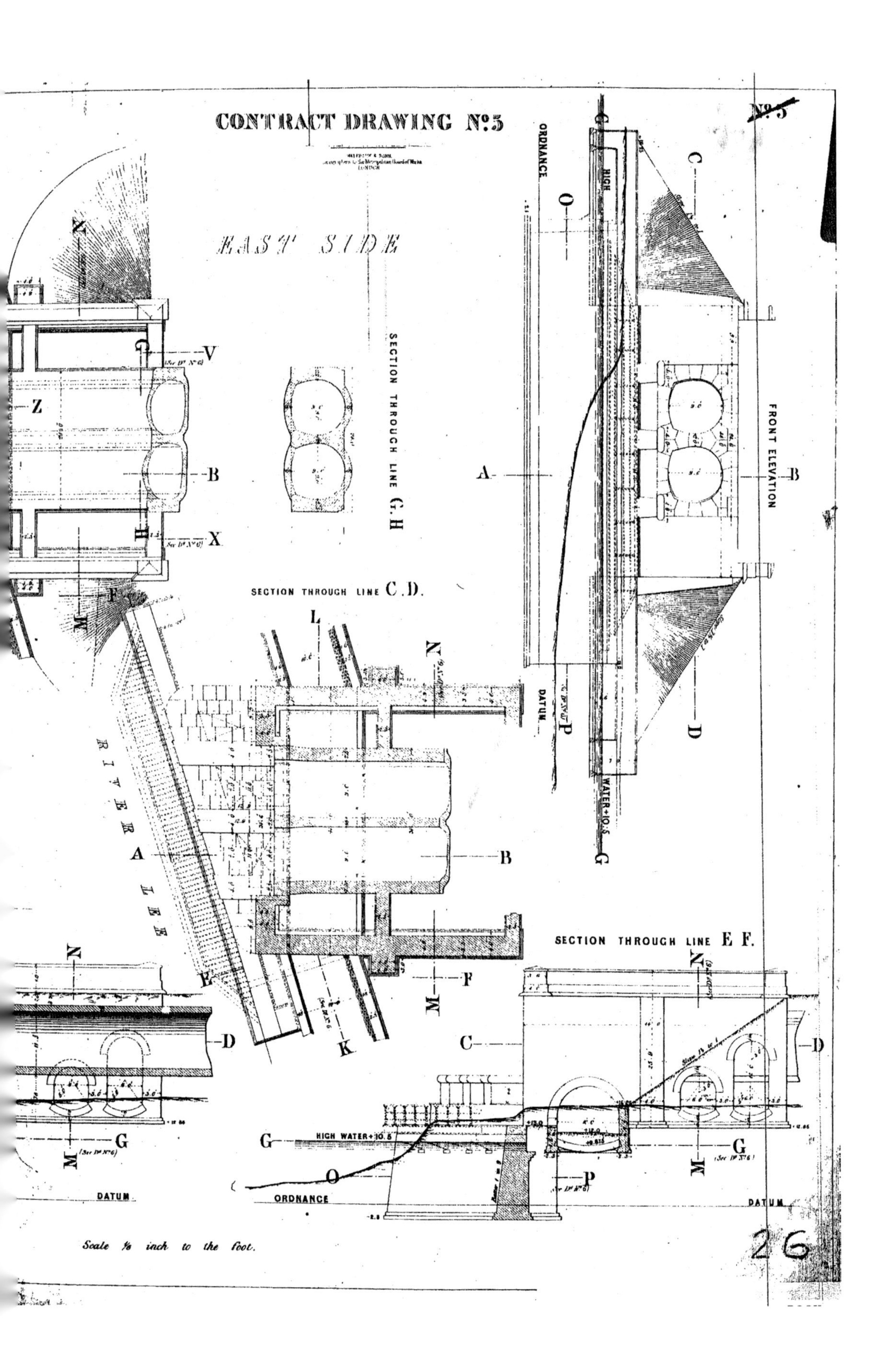
CONTRACT DRAWING No.5
EAST SIDE
SECTION THROUGH LINE G.H
SECTION THROUGH LINE C.D.
FRONT ELEVATION
ORDNANCE
HIGH
DATUM
WATER +10.5
RIVER LEE
SECTION THROUGH LINE E F.
HIGH WATER +10.5
DATUM
ORDNANCE
DATUM
Scale ½ inch to the foot.
26

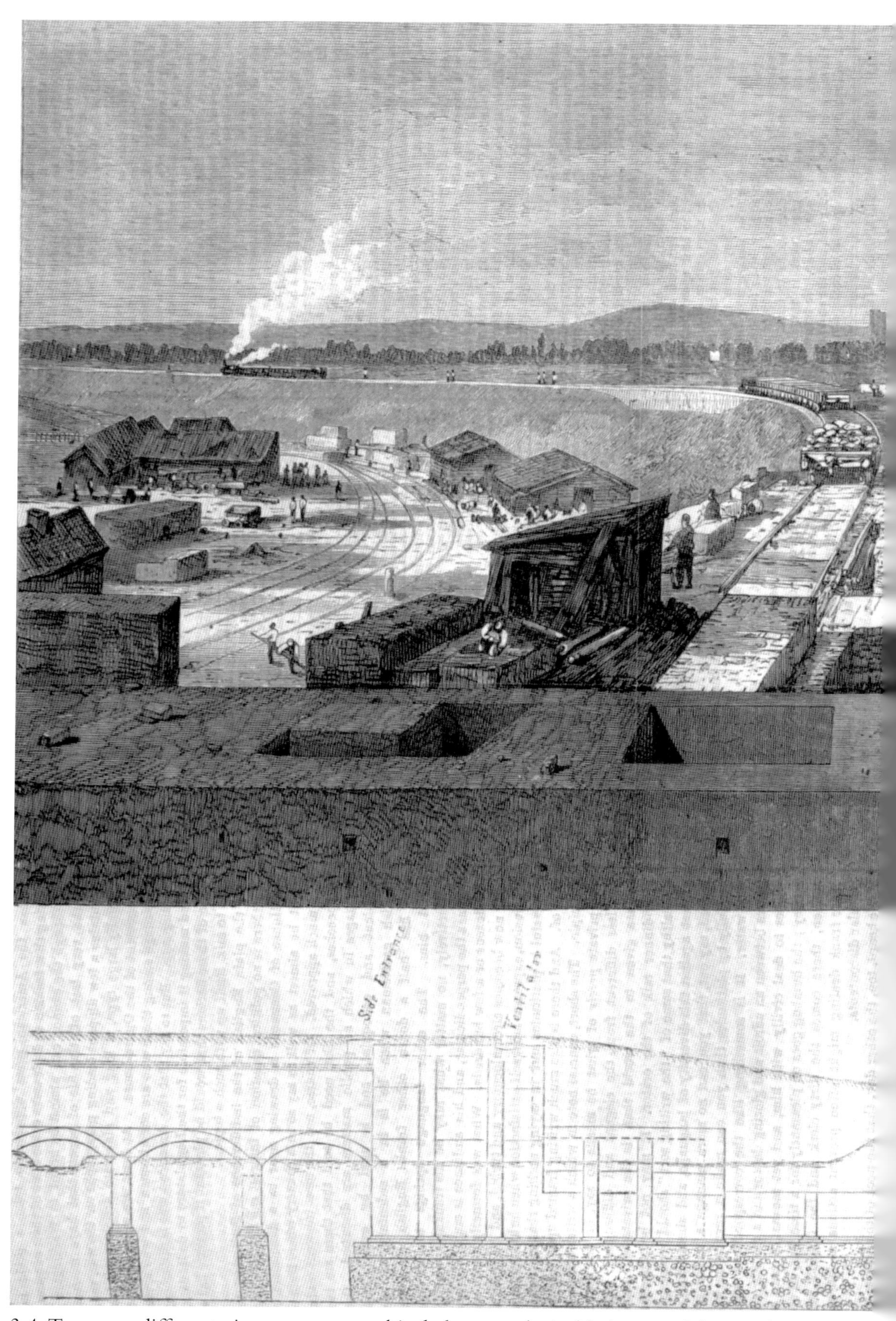

3.4 Two very different views – topographical above, technical below – of the northern outfall sewer in the *Illustrated London News*, 28 May 1864, p. 513.

TRINITY HIGH WATER

original orthographic context: that is, unlike the original contract drawing, it does not show a projected view of image directly above it. Rather, the *Illustrated London News* itself defines the 'technical' meaning of this drawing; unlike the *Engineer*, this newspaper is not addressing a specialist audience; rather a multi-faceted group of middle-class Londoners with a range of technical abilities. As a composite image – at once both dramatic and technical – this engraving serves to both educate and engage these readers in a way that would have meant very little to an engineer or contractor.

Specification

If the northern outfall sewer contract drawings can be said to have many different functions and attendant audiences, then the accompanying detailed written specification might be expected to be directed at one audience only: the range of bidding contractors and subsequently the successful bidder, in this case George Furness. The specification, read in conjunction with the contract drawings, provided the prospective contractor with much more complete information than the drawings alone, as well as articulating more fully his projected relation with Bazalgette. The form and scale of the main drainage specifications varied from contract to contract: for example, the three contracts produced for the northern outfall project include an 1859 contract for iron fencing made up of a four-page hand-written specification and a sketch drawing;[38] while the 1863 contract for the outfall reservoir comprised 32 pages, 128 clauses and ten detailed drawings.[39] However, the huge scale of the 1860 northern outfall sewer contract set it apart as an engineering project on a par with the building of an extensive line of railway: it consisted of 85 printed pages, besides a 6-page index listing the 267 separate clauses.

The northern outfall sewer specification is divided into two parts: first, a series of clauses giving detailed instructions to the contractor; and second, a list of quantities and prices of materials required.[40] The order of the clauses follows that of the drawings, with the earlier clauses outlining general instructions regarding construction (clauses 1-17), the later detailing the specific design features such as the proposed aqueduct over the River Lee (clauses 18-50). Where the drawings show many detailed visualisations of the project, both general and specific, the specification directly instructs the contractor as to his duties. The information given in the clauses often refers directly to the contract drawings: seven clauses (numbers 4-7 and 12-14) in the index refer to drawing 47; while those outlining the arrangements for the aqueduct over the River Lee begin with general references to the drawings (clauses 18-20 refer to drawings 5-8) and end with references to specific parts of individual drawings (clauses 20-49 refer to drawings 9-12).

The last pages of the specification comprise a schedule of prices, fixed by Bazalgette's estimators in advance and common to all the main drainage contracts. A quantity surveyor, working from Bazalgette's contract drawings, calculated the volumes of materials required and their respective costs.[41] The resulting exhaustive list was then incorporated into the specification, including estimated prices for all materials – per cubic yard for concrete and per 'rod' for bricks – and all types of labour, including two types of digging (general excavation and foundation work) – paid per cubic yard excavated – and more specified types of labour paid on the basis of a ten-hour day's work.[42] The contractor would then be expected to produce his own estimates that had to correspond with the prices set by the engineer. Bazalgette ordered that the entire project was to be completed in two years; a fine of £50 per day would be levied on the contractor thereafter.[43] The contractor would then maintain the works for one year after completion after which he would receive his full payment.[44] For the duration of construction the contractor was paid on a monthly basis, with 90 percent of the contractor's expenses being paid during this time; five percent paid on completion; the remaining five percent paid after the period of maintenance.[45] By specifying all of these details in advance, Bazalgette ensured that all bidders would conceive of the project along the same lines, so that their bids would be comparable. In effect, by relieving the contractor of the risk of underestimating the cost of materials, the resulting range of bids would say much more about the projected efficiency of organisation of each of the contractors, enabling Bazalgette and the Board of Works to ensure, in their final choice of contractor, both cost-effective and quality construction.

Throughout the specification, the drawings figure as important, yet incomplete, mediating representations between engineer and contractor. The contractor was ordered 'to set out the works in every particular, according to the drawings', with 'the written dimensions on the drawings ... in all cases to be preferred to the scales, in computing the quantities'.[46] Yet this unequivocal assertion of the primacy of the drawings is also undercut in one of the introductory clauses:

> Such drawings and specification are to be considered as explanatory of each other, and should anything appear in the one that is not described in the other, no advantage shall be taken of any such omission, and the Contractor is so to consider in forming his estimate. Should any discrepancies, however, appear, or should any misunderstanding arise as to the meaning and import of the said specification or drawings ... thereupon the same shall be explained by the Engineer for the time being of the Metropolitan Board of Works, and this explanation shall be final and binding upon the Contractor; and the Contractor is so to execute the work according to such explanation, and without charge or deduction to or from the Contract.[47]

Here, Bazalgette outlines the co-dependency of the drawings and specification and also his own responsibility for their correctness or otherwise. Although bound to any decisions made by the engineer to alter either the drawings or specification, the contractor clearly has some degree of freedom to question the meaning of the specification or drawings, the quality or dimensions of the materials, or what is meant by the 'due and proper' execution of the works. Furthermore, with regard to some information depicted in the drawings – levels and borings – it is the contractor, rather than the engineer, who is deemed responsible for any errors or omissions. The fact that the contractor 'must satisfy himself as to the correctness of this information' points to his level of freedom in providing his own measurements to carry out construction.[48] As will be seen below, this ambiguity was to result in conflicts during construction between engineer and contractor, centred on the payments made from the Metropolitan Board of Works to the contractor.

Construction

Whereas early Victorian engineers have received detailed attention from biographers, the men who founded the large-scale contracting companies in this period, such as William Webster, Thomas Brassey, John Aird, and George Furness, have earned relatively scant recognition.[49] This neglect has been attributed to the fact that unlike engineers, who left voluminous verbal and visual documentation, contractors left few records as their business was mostly carried out orally.[50] The official record of the construction process comes from Bazalgette himself, who was asked by the Metropolitan Board of Works to produce monthly reports throughout its duration,[51] in addition to his accounts already included in the annual reports of the Board.[52] Apart from press accounts, Bazalgette's reports remain the only thoroughgoing documentation of the construction process. However, the facts and figures that make up the bulk of his reports do not provide much insight into how these statistics were compiled. Throughout the construction process, important parts of Bazalgette's engineering department were the site-based staff – including surveyors, 59 clerks of works and his three assistant engineers (the latter supervised construction). The clerks of works were the most important mediators between contractor and engineer and were permanent residents on site for the duration of construction. Their duties included measuring the contractor's work on a weekly basis, regulating construction lines and levels, and assuring the quality of materials used.[53] In effect, they provided the raw material for Bazalgette's monthly reports.

Bazalgette's monthly report of February 1861 stated that the line of the northern outfall sewer had been marked out with fencing and that construction had begun soon after.[54] Furness, like many other large-scale contractors, set up his own concrete mixing works on site that included five stationary engines, freight cars and a specially built line of track leading from Barking Creek to the site. Bazalgette stated in June 1861 that these initiatives 'inspire one with confident expectations'.[55] By August 1861, Furness began constructing the brickwork and aqueduct over the River Lee, as well as many of the other aqueducts along the route and the concrete foundations and brick arches for the embankment across the marshes.[56] By January 1862, the iron girders and superstructure for the aqueducts were being manufactured in off-site foundries.[57] In February, Furness's men had cut through the embankment of the Eastern Counties Railway to a depth of 24-feet 'without interruption to the traffic'[58] and by October, four miles of the concrete embankment were completed.[59] By the middle of 1863, the brickwork for the double and treble lines of sewer was almost complete with all the ironwork for the aqueducts on-site and ready to be erected.[60] By August, Bazalgette stated that the project was 'near completion',[61] although it was not until March 1864 that he was able to say that it had been finished.[62] Work on the outfall and reservoir at Barking Creek, also carried out by Furness, continued until August 1864.[63]

Bazalgette's reports present the construction of the northern outfall sewer as a relatively seamless process, unimpeded by contingencies: in effect, merely a vast accumulation of materials and money spent. When we do 'hear' the contractor's voice, such as in a report Bazalgette compiled to justify a 30 percent increase in his original cost estimates, information given by the contractor is used solely to legitimize Bazalgette's argument.[64] Likewise other contingencies – the 1859-60 labourers' strike over working hours, continuous problems with heavy summer rains, hard winter frosts, and difficulties obtaining enough brick and stone – are all presented as factors that, although they often delayed construction, did not prevent progress remaining 'decidedly satisfactory'.[65] However, even a cursory examination of the minutes of the meetings of the Metropolitan Board of Works in the early 1860s shows that Bazalgette conceals much in his account – hardly surprising given that he was presenting his reports to the same audience that had approved his initial cost estimates. The Board minutes are filled with claims from the various contractors that questioned Bazalgette's calculations. For example, in 1865, after completing the northern outfall sewer contract, Furness was still claiming outstanding payments that contradicted Bazalgette's initial estimates.

According to Bazalgette, in such an event, the relevant assistant enginer and the clerk of works would discuss the issue with the contractor and his agents using the contract drawings and specification as a guide.[66] The correct costs would be determined and a list sent to the accountant who would, by comparing it to the engineer's schedule of prices, certify and sign it as correct. However, because Furness was disputing the initial cost estimates themselves – that is, those made on the basis of the information in the drawings and specification – this process was clearly an inadequate way of resolving the problem. After receiving repeated letters from Furness demanding to be paid, Bazalgette eventually suggested appointing an independent engineer to arbitrate.[67] The Metropolitan Board of Works firmly rejected this proposal stating that it had 'the fullest confidence in their engineer'.[68] The outcome of this apparent impasse is not known. Because Bazalgette had already provided Furness with a degree of flexibility in his interpretation of both drawings and specification, he was clearly torn between maintaining his alliance with Furness and the procedures approved by the Board. Bazalgette's decision to absolve himself of responsibility is tantamount to an admission of defeat: a decision revoked by the Board, who obviously understood the possible implications of being seen to undermine the confidence they placed in their chief engineer.

Bazalgette's written accounts of the construction of the main drainage system were given a pictorial counterpart in a series of photographs commissioned by the Board of Works in the early 1860s.[69] These images documented moments when the main drainage system was opened up to the scrutiny of an 'official' public; they depict visits to the construction site organised by the Board in

3.5 Photograph showing the construction of the northern outfall sewer in 1861 depicting in the bottom-right foreground a contract drawing being held by Bazalgette, George Furness and an unidentified figure.

October 1861, July 1862 and July 1864, attended by representatives of London's newspapers, vestrymen, and members of the Lords and Commons respectively; and lavish ceremonies held to mark the completion of the Crossness pumping station in April 1865, attended by royalty, Members of Parliaments, Archbishops and numerous other dignitaries. The Board of Works set a precedent in its use of photography to document the construction of high-profile technological projects.[70]

The photograph reproduced in **3.5** shows a portion of the northern outfall sewer under construction from 1862; prominent, on the left, is one of Furness's concrete making machines, developed specifically for this project. In the background, on the right, the semicircular forms of the embanked sewers can be seen, with two wooden pulley devices in front. Scattered throughout the image are various workmen – the supervisory staff distinguished by their dark clothing and top hats, the manual labourers mostly carrying shovels and buckets. However, the principal subjects of the photograph are the three figures standing in the mud at the bottom centre-right of the image gazing towards the camera. Bazalgette is the only immediately recognisable figure on the left; the middle figure is probably Furness, while that on the right is likely to be either the clerk of works or the on-site engineer, Edmund Cooper. A white piece of paper, almost certainly one of the contract drawings, is held by all three figures, serving to link together their static poses.

This photograph is unique among the collection commissioned by the Metropolitan Board of Works in depicting a contract drawing. Here is an immediate visual representation of the alliance between engineer, his associates and the contractor, with the drawing cementing this relationship. The image also presents the viewer with a correlation between this relationship and the reality of construction seen to the left and behind the figures, with its welter of hastily constructed wooden supports, its array of scattered workers, the prominent mounds of displaced mud and dirty trenches, and what appears to be a discarded overcoat directly behind the figure of Bazalgette. If the photograph confirms the importance of the alliance of Bazalgette and Furness it also raises questions as to just how they negotiated the variety of other figures seen throughout the image: from the supervisors of the works, in their top hats and black clothing, to the labourers, or 'navvies', seen in the background right of the image, inside the concrete making machine, and, prominently in the left foreground, passively gazing at the camera. How did Furness manage such an enormous workforce? What role did the drawings play in this process? What was the relationship between Bazalgette and his variety of supervisory staff and, in turn, between them and the navvies? The absence of written records on the part of Furness coupled with the dominance of Bazalgette's representational bias makes it very difficult to gauge the precise nature of these interactions. What this photograph provides is, for a brief but permanently captured moment, a reflection on the range of voices and the complexity of relationships missing from the existing documentation. The next chapter will focus more squarely on some of these other voices: namely, those of the press, who, in rapt attention, provided a very different picture of the construction process than Bazalgette's representations.

4

Sublime Spaces

During the 1860s London's infrastructure was physically transformed. The building of the main drainage system was but one gigantic work among many begun in this decade, including the Metropolitan Underground Railway (from 1860), the Thames Embankment, containing part of the main drainage system (from 1862), the London, Chatham and Dover Railway (also from 1862), and new street improvements such as the Holborn Viaduct (1866-69). Such building works brought chaos to the streets of the city as well as new spectacles of both excavation and ruination, translated into a wide variety of imagery that filled the pages of London's illustrated press. Victorian Londoners had never before experienced construction on this industrial scale: the sheer vastness of the spectacle overawed and, at times, overwhelmed the public; and there was one concept in particular that seemed to be able to represent these feelings: the sublime.

This chapter considers press responses to the construction of the main drainage system in the early 1860s, focusing on wood engravings in the *Illustrated London News*, which gave by far the most extensive visual coverage of the project. As will become clear, these press responses differ sharply with Bazalgette's presentation of his sewers and their construction, considered in chapter 3. In the press accounts and engravings, the sewers can no longer be seen as purely instrumental spaces; rather, they are experienced as real spaces in the city: a visceral spectacle provoking both wonder and anxiety. If the new sewers carried with them the idea of heroic progress, they were also, at times, seen as a force of destructive power; if the new works were comparable to the ancient wonders of the world, they also contained within them the sense of demonic, subterranean forces out-of-control.

The Industrial Sublime

The concept of the sublime, popularised in the mid-18th century by writers and theorists such as Edmund Burke (1729-97), had, by the early-19th century, become established as an important artistic trope.[1] Defined by Burke as a strong emotional response – made up of a mixture of awe and terror – to vast or overwhelming natural or man-made objects,[2] the sublime readily appealed to

Left: 4.1(a) Frederick Smyth's wood engraving of the deepening of the Fleet-street sewer in the *Illustrated London News*, 4 October 1845, p. 213.

Above: 4.1(b) Frederick Napoleon Shepherd's watercolour sketch of the same scene in 1845 on which the engraving was probably based.

artists who wanted to move the spectator in ways that the existing aesthetic categories of the beautiful and the picturesque did not.[3] For the British painter J.M.W. Turner (1775-1851), primal nature provided the dominant subject matter in canvases that gave visual form to Burke's categories of the sublime, such as 'Vastness', 'Obscurity' and 'Power'.[4] In some of his later works, new industrial spectacles, such as a train speeding through the countryside in *Rain, Steam and Speed – The Great Western Railway* (1844), also provided an equivalent sense of sublimity. In the contemporaneous art of John Martin (1789-1854), both natural and industrial forms provided equal inspiration; his sensationalist imagery sometimes borrowed directly from contemporaneous 'sublime' urban industrial spectacles, most notably the Thames Tunnel in his 1827 mezzotint *The Bridge Over Chaos*.[5] By mid-century the identification of the sublime with new technological forms, such as railways, bridges and factories, formed a key component in the presentation and reception of industrial development in both urban and rural contexts.[6]

If the industrial sublime was an important theme for artists such as Turner and Martin, its appropriation to the more mundane medium of wood engraving was informed by the ideological leanings of its main vehicle, the illustrated newspapers. From 1832 onwards, with the publication of the *Penny Magazine*, wood engraving was promoted as a medium ideally suited to illustrating mass-produced newspapers, made possible by improvements in the mechanisation of the printing process, the reproductive longevity of wood engravings, and the ease with which they could be printed alongside text. Whereas the *Penny Magazine* concentrated on the production of engraved versions of well-known works of art in order to elevate working-class taste, the *Illustrated London News*, founded in 1842, marketed itself specifically as an illustrated newspaper and, initially costing sixpence, appealed rather to a 'respectable' middle-class audience.[7] This aim of conservative respectability set the *Illustrated London News* apart from other weekly newspapers in London which tended to be politically left-of-centre, but it did not prevent the newspaper from becoming enormously successful: by the late 1850s it had higher circulation figures than any of its rivals, selling as many as 100,000 copies every week at a reduced price of five pence. Such developments stimulated an enormous growth in the wood-engraving trade in London even before rival illustrated newspapers challenged the dominance of the *Illustrated London News*.[8]

From the start, the *Illustrated London News* adopted a high moral tone with regard to the subject matter it would embrace – in effect, its goal was to invest the everyday with a sense of sublimity normally associated with the 'high' art of painters like Turner or Martin. On 27 May 1843, on the occasion of its first anniversary, the newspaper stated that its founding aim was:

> … to plunge into the great ocean of human affairs, and to employ the pencil and burin in the work of illustrating not only the occurrences of the day, but the affections, the passions, the desires of men, and the faculties of the immortal soul.[9]

Such lofty ideals reflected the desire of the newspaper's editors to elevate both the value of 'news' events and the medium by which these events would be imaged – that is, wood engraving. In effect, it aspired to provide an elevating record of historical events – a role more characteristic of history painting in the Royal Academy.[10] This aim – to simultaneously elevate the public appreciation of news events and wood engraving – was embodied in the way in which the newspaper appropriated the notion of the sublime in its representations of the most everyday news events, such as the building of sewers. From the outset, the *Illustrated London News* had set itself up as a mouthpiece in the campaign for metropolitan 'improvements' and, in particular, consistently voiced the need for reform in the city's sanitary infrastructure. In 1845, the sewer in Fleet Street was excavated by the Holborn and Finsbury Sewer Commission and was deepened and enlarged to improve its efficiency and capacity. The *Illustrated London News* published a report on the work including two illustrations.[11] **4.1(a)** shows one of these: an engraving of the interior of the sewer during excavation. The visual impact of the illustration mirrors the tone of the article, which celebrates this improvement and the 'idea of the extraordinary labour' required to excavate and deepen the sewer.[12] By comparing this engraving to a contemporaneous watercolour by Frederick Napoleon Shepherd depicting the same scene (**4.1: b**), the sense of exaggerated scale in the former can be appreciated: the narrowing of the composition, the addition of several more wooden struts, and the inclusion of two diminutive workers at the bottom of the trench serve to magnify and impress upon the viewer the scale of the works.[13] Even the precise, linear technique of the engraved image contributes to this by clearly articulating the forms of the excavation in a way that the watercolour does not.

The sense of exaggerated scale in the engraving corresponds to what some scholars have described as the 'rhetorical' character of visual depictions of urban industrial forms in this period: that is, images of industry designed to have a deliberately persuasive or impressive effect. Nicholas Taylor argues that such rhetoric was key to the 'emotional appeal embedded in the sublime'.[14] Likewise, Rosalind Williams sees the vocabulary of the sublime as 'a rhetorical mode' that was employed in Victorian representations of industry and directed at particular groups of people who were disturbed by the disruption caused by urban technological change.[15] Such rhetoric was bound up with the promotion of middle-class notions of industry as progressive and key to the improvement of the urban environment.[16] Those who promoted such an ideology of improvement readily drew upon aspects of the sublime that would serve to heighten this sense of nobility. Therefore, in relation to the image of the Fleet sewer, one can read the exaggerated scale of the excavation as a means by which the newspapers' readers (upper middle-class Londoners) might be persuaded of the benefits of sanitary improvement – readers who may have resented the disruption the excavation undoubtedly caused.

However, the readers of the *Illustrated London News* would not have viewed this dramatic image in isolation; rather, in the context of a page containing both text and other images (**4.2**). The engraving is importantly one of a pair: the image at the bottom right of the page, also similar to another watercolour and sketch by Frederick and Thomas Shepherd, depicts the same scene as that in **4.1(a)** but from above ground.[17] Such exterior/interior image combinations were a common device employed by the newspaper to comprehensively describe particular scenes or events to its readers; this prompts a modified reading of the Fleet sewer image – one that takes into account the important function of wood engraving in the *Illustrated London News* as a realist medium.[18] If illustrated newspapers drew on notions of the sublime as a way of imbuing their engravings with emotion and drama, they also promoted the same engravings as a form of documentary representation.[19] From its founding in 1842, the editors of the *Illustrated London News* consistently stated their intention not only to elevate the status of wood engraving to that of a fine art but also to use the medium to realistically document the most everyday of events, such as the building of sewers. These twin aims come together in the engraving of the Fleet sewer – an image that is at once dramatic and descriptive. If the clean lines of the engraving bring out the sublime scale of the spectacle, they are also a precise form of documentation intended to educate the viewer.

The Celebratory Sublime

Whereas the deepening of the Fleet sewer was a localised event causing only small-scale disruption, the construction of the main drainage system in the 1860s was a vast citywide undertaking: 82 miles of new main sewers were built, running in parallel lines from west to east across London, made up of 318,000,000 bricks, 880,000 cubic yards of concrete, and requiring the excavation of 3,500,000 cubic yards of earth.[20] The scale of the project, despite causing considerable disruption to the urban fabric, was consistently extolled by the press. In January 1859, at the start of construction, the *Illustrated London News* was already ascribing a sublime status to the project: 'in the course of the coming summer this drainage scheme – of a magnitude unknown, at least in modern ages – will have become a work of general interest to ourselves and to the numerous foreigners who may visit our shores'.[21] The works are seen as comparable only to the engineering feats of the ancient world, its 'wonders' now to be given a modern equivalent. Such ascribing of a mythical status to the project was to become a common rhetorical device used by the press to convey the vast size of the largely underground works and their 'nobility' as essential metropolitan improvements.

The first large-scale representation of the project in the *Illustrated London News* gives visual expression to this rhetoric. In August 1859, the newspaper published a full-page article describing the progress of the works and illustrated by a large engraving showing the building of the sewers at Wick Lane, near Old Ford in Hackney (**4.3: a**). The article states that the engraving was produced 'from a photograph by Mr. F. Thompson' and represents the point at which two of the new sewers (the northern middle and high level sewers) joined.[22] **4.3(b)** shows a photograph depicting the same scene as the engraving and, although not attributed, it is likely that this photograph, or another of similar composition, formed the basis for the newspaper's engraving.[23] However, the compositional elements of the engraving differ significantly from those of

Oct. 4, 1845.] THE ILLUSTRATED LONDON NEWS. 213

THE FLEET-STREET SEWER.

The works in progress for deepening the Sewer of Fleet-street have attracted considerable attention, partially from the obstruction which they have presented to the public traffic. They are, however, of intrinsic interest; although the Sewer of Fleet-street cannot compete, in antiquity, with the ancient water-course known as the Fleet Ditch,—

> The king of dykes, than whom no sluice of mud
> With deeper sable blots the silver flood.

It appears that although Sewers have been constructed in London for upwards of four centuries, it is only within the last ten or fifteen years that the drainage of the City has been satisfactorily accomplished. Hitherto, it was very defective and imperfect; some of the smaller streets having a Sewer, while the larger thoroughfares, as Cheapside, Ludgate Hill, &c., had none. From time to time, however, this evil has been remedied; and the Sewerage is now nearly completed.

The Sewer of Fleet-street, the subject of our Engraving, having been found insufficient to carry off the water, for which it was intended, it became necessary to cut deeper, and construct a new sewer: the greatest requisite depth is 25 feet, which decreases to about 17 feet near Temple Bar; hence the Sewer runs easterly to Water Lane, where it is joined by another Sewer, which runs into Whitefriars Dock.

One of our Engravings conveys an idea of the extraordinary labour requisite for excavating the ground to the requisite depth, and the numerous provisions against accidents in the dangerous operation. Such is the underground labour; whilst the difficulty of keeping open the traffic, so as not to extinguish the "very animated appearance" of Fleet-street, is a work of much difficulty. The cost of the present undertaking, contracted for by Messrs. Ward and Son, of Aldersgate-street, is £2000.

There do not appear to be published data from which the total extent of the metropolitan Sewers can be ascertained. The Holborn and Finsbury divisions contain eighty three miles. In addition to these, there are sixteen miles of smaller Sewers, to carry off the surface-water from the streets and roads, and two hundred and fifty-four miles of drains leading from houses to the main Sewers.

THE NEW SHERIFFS.—The new Sheriffs of London and Middlesex, Mr. William James Chaplin and Mr. John Laurie, were sworn into office on Saturday last, at Guildhall. Mr. David W. Wire is appointed Under-Sheriff to Mr. Sheriff Laurie; and Mr. C. Burcham, of Red-lion-square, Under-Sheriff to Mr. Sheriff Chaplin. A vote of thanks was unanimously passed to the late Sheriffs, Alderman Hunter and Alderman Sidney.

NARROW ESCAPE OF THE MAYOR AND CORPORATION OF LONDON.—The *Times* has the following pathetic account of the narrow escape of the Mayor and Corporation of London:—"The worthy citizens of London who are peacefully following their various avocations, have, we fear, only a faint idea of the labours and perils which the members of the Corporation are constantly exposed to in the execution of their arduous duties. The passage of the river from London-bridge to Chelsea is now so well known to the inhabitants of London; the 'move her a-head,' and 'turn her a-starn,' are phrases so common in our ears, and convey such a perfect idea of the control and safety which pervade throughout the navigation of the river, that few will believe that the lives of the whole Corporation were, on Tuesday, nearly sacrificed through the perils of transit by the City barge, from Blackfriars to Westminster. We will not venture to say to whose pilotage this catastrophe is to be ascribed, but we apprehend that the attention of the Aldermen has been recently so much devoted to the management of railways, that they have lost sight of those ancient land-marks by which their official and swan-hopping excursions have been heretofore guided. The consternation which prevailed after the alarming accident which we are about to detail, was such that we are unable to fix accurately the time at which it took place, but we are enabled to state that the Lord Mayor was in the cabin, seated in his chair of state—the mace was on the table before him—the table also contained the usual means of refreshment, and was surrounded by some of the Aldermen, the late Sheriffs, and the Sheriffs elect, when, almost simultaneously, with a cry of 'A-head, there!' the City barque came in violent collision with one of the piers of Westminster-bridge. The Lord Mayor, and, we believe, nearly all those present, were thrown from their seats by the violence of the shock. The mace, the decanters, the glasses, and the wine rolled from the table, and the members of the Corporation were prostrate in one common wreck on the floor of their barge. Happily, the alarm subsided with the shock, and, with the exception of a few bruises, we have the satisfaction to state that the worthy members of the Corporation escaped unhurt."

INAUGURATION OF THE STATUE OF JEAN BART, AT DUNKERQUE.

(*From a Correspondent.*)

A *fête*, upon a grand and extensive scale, took place last month, at Dunkerque, on the Inauguration of the Statue of Jean Bart, a town-born naval hero of the place. For a month previous, the town had been under great excitement; and no expense was spared to have the ceremony conducted with an *éclat* and splendour worthy of the occasion. Invitations were sent to twelve of the principal towns of the department, including Calais and St. Omer, to have the attendance of the musical bands of their respective National Guards, which were readily accepted; and from eight to eleven o'clock on the day of the Inauguration, were seen marching into the town, the different bands playing national airs, at the head of from 50 to 100 men of their respective corps. These formed a circle around the veiled Statue, and composed a musical force, united with the military bands of the town, of nearly 600 instruments. Precisely at twelve, the veil from the Statue dropt, and discovered the hero, in the costume of the time of Louis XIV., brandishing a sword in his right hand, a pistol in his left, and wearing another in his belt. He is supposed to be engaged in boarding a ship, as this was his usual and favourite mode of carrying on the fight, and which ensured to him so much success: he succeeded in capturing and carrying into Dunkerque upwards of two hundred Dutch merchant vessels, during the war with the Dutch States. The Statue is of bronze, 16 feet high, and executed in the first style of the celebrated David (d'Angers)—the Thorwaldsen of France.

The unveiling of the Statue was hailed by the acclamations of 40,000 voices, and followed by a crash of 600 instruments. Orations were then delivered by the Sous-Préfet, the Mayor, and others, enumerating the numerous exploits at sea performed by this heroic sailor, and holding him up as an example for others to emulate. In the afternoon, all sorts of amusements, both on water and land, were provided for the populace, and the evening concluded with balls, concerts, &c. (gratis). A general illumination followed; fireworks and balloon-flying finished the day.

The following day the pastimes of the people continued, though diversified; but the most interesting sight was a dinner party of 1800 persons. Four tables were laid out in the grand avenue of the park, and at one o'clock marched in the bands of music and corps of the different towns of the department, and regularly took their seats assigned to them: this was known by a tally-flag affixed to each bottle of wine, and every man found by the side of his plate a bottle of claret. The dinner consisted of cold joints, ham, meat pies, &c. The wine was excellent, and toasts were drunk of such a patriotic character, and of so much enthusiasm, that every man seemed to consider himself a second Jean Bart. Fireworks, the ascent of a balloon in the Grand Place, and a general illumination, closed the celebration of the Inauguration of the Statue of their valiant townsman.

STATUE OF JEAN BART, AT DUNKERQUE.

The decorations of the houses and streets were unique and tasteful. Festoons of drapery of different colours were stretched from house to house; while from the windows floated tri-coloured flags, banners, bearing insignia or inscriptions of the exploits of Jean Bart. Triumphal arches spanned all the streets, while each side of them was lined by rows of trees, or large branches, resembling young elms, of four years' growth. The transformation of these streets into groves, with triumphal arches, covered with garlands and trophies, and which terminated the "vista," had a most singular and beautiful appearance.

FLEET-STREET.—DEEPENING THE SEWER.

4.2 Page layout from the *Illustrated London News*, 4 October 1845, p. 213 showing two views of the excavation of the Fleet-street sewer.

4.3(a) A sectional view of the tunnels of the northern outfall sewer at Wick lane in Bow in the *Illustrated London News,* 27 August 1859, p. 203.

4.3(b) A photograph depicting the same scene in 1859 on which the engraving was probably based.

August 27, 1859.] THE ILLUSTRATED LONDON NEWS 203

MAIN DRAINAGE OF THE METROPOLIS.

All those persons who frequent this great city, and have, during the last month or two, experienced the evils arising from the present condition of the Thames, cannot fail to be interested in the progress of the works which are now being constructed in order to purify its waters.

It is quite true that London is the best drained city in the world, and that the construction of from four to five hundred miles of sewers within the last seven or eight years, in addition to the thousand miles of sewers which previously existed, has tended greatly to improve the sanitary condition of our back streets and by-ways, which form the haunts of the poorer classes, and it is estimated that 200,000 cesspools have of late years been removed from beneath our dwellings, and thus many houses that were previously scarcely habitable have been rendered wholesome. Thus far we have done well, and it would appear that we have at the same time been fully alive to the fact that these sanatary operations tended to the pollution of our noble river, for all have acknowledged the necessity for some great scheme which should relieve poor Father Thames of the humiliating duty which he is now required to perform; but, unfortunately, so many schemes were propounded, that year after year was wasted in controversy, and nothing was done. Indeed, we may fairly assume that if the river had not itself cried loudly for relief, we might still have been debating as to the means by which the cure could most advantageously have been effected, instead of progressing as we now are with remedial measures.

It was about this time last year that Parliament placed in the hands of the Metropolitan Board of Works the means for carrying out the plans which were designed by Mr. Bazalgette, their engineer, in the year 1854, and which have since that period been open to the criticism of the engineering profession and the public generally, and the details of which have thus, from time to time, been modified and improved.

Mr. Bazalgette's scheme proposes, by constructing three arterial lines of sewers on the north side of the Thames, converging to one point at the River Lea, and passing thence side by side in one large embankment to Barking Creek, to divert the sewage from the Thames; and, should it not be possible to make it available for agriculture, to cast it away at a point fourteen miles below London-bridge, at high water, so that the ebb tide may carry it twelve miles lower, or twenty-six miles below London-bridge; and similarly to deal with the sewage on the south side of the river.

The three lines on the north side he calls the high-level, the middle-level, and the low-level sewers. The two former will discharge their waters by gravitation, and the contents of the latter will be raised by steam-engines and pumps.

Early in the present year Mr. Moxon obtained the contract for the high-level sewer, at a price considerably within Mr. Bazalgette's estimate; and the rapidity with which the works have been conducted, as well as their substantial and lasting quality, are equally creditable to Mr. Moxon and to Mr. Cooper, the resident engineer. The work already executed is valued at over £50,000, and includes the construction of about 9000 feet of sewer or tunnel, varying in size from ten to twelve feet in diameter, 2000 rods of brickwork, 200,000 cubic yards of excavation, and 10,000 cubic yards of concrete, and it gives employment to upwards of two thousand artificers and labourers.

These tunnels being generally underground, it is not often that an opportunity is afforded of presenting a view of the works, which are now opened in various localities between the River Lea and Hackney. Our Engraving (from a photograph by Mr. F. Thompson) represents the works immediately beyond the junction of the high and middle level gravitating sewers at a point between the North London Railway and the River Lea, near Old Ford, Bow, where these sewers emerge from high ground into the valley of the Lea, preparatory to being carried over that river by an iron aqueduct, and by an embankment across the West Ham and Barking marshes. Each of these tunnels is large enough to carry off the maximum flow of sewage which will arise from the calculated prospective increase of the population of London many years hence, together with a fall of rain upon the area drained equal to a quarter of an inch deep in twenty-four hours, in addition to that which is absorbed and evaporated; but each sewer is also provided with a storm-overflow or safety-channel underneath it, so that on occasions of heavy rains and thunderstorms, the surplus will flow over weirs provided for the purpose (into and through the storm channels) into the River Lea.

BLACK SWANS AND THEIR YOUNG, AT CULVERS, SURREY, THE SEAT OF S. GURNEY, ESQ., M.P.—SEE SUPPLEMENT, PAGE 216.

MAIN DRAINAGE OF THE METROPOLIS.—SECTIONAL VIEW OF THE TUNNELS FROM WICK LANE, NEAR OLD FORD, BOW, LOOKING WESTWARD.

4.4 Page layout from the the *llustrated London News*, 27 August 1859, p. 203, showing the dominance of pictures over text in the newspaper.

the photograph: the viewpoint, slightly raised and set back to the left of the works in the photograph, is positioned at the very bottom of the trench, closer and square-on to the works in progress. This alteration allows the scene of construction to dominate, excluding most of the distant features shown in the photograph; one tree remains at the top right of the engraving, forming an odd juxtaposition with the adjacent gigantic forms of the sewers. In short, the close-up view in the engraving serves to exaggerate the scale of the works. Such compositional elements have the effect of magnifying the scene of construction in both its scale and proportion to the viewer. Like the earlier image of the Fleet sewer, this engraving can be read as an embodiment of the sublime – that is, a deliberately dramatic representation of a spectacle that would prompt a celebratory, awe-struck response.

Focusing now on the arrangement of the image on the page (**4.4**), I want to bring out some important differences between this engraving and the earlier images of the Fleet sewer (**4.2**). During the 1850s, the *Illustrated London News* vastly expanded both the number and size of its engravings in individual issues so that, by 1859, the engravings dominate the page with the text being squeezed into the smaller available spaces. Such a change significantly alters the relationship of text to engraving, reinforcing the dominance of the image and consequently its dramatic possibilities. Larger engravings in the newspaper were also generally made differently from their smaller counterparts: a single known artist, Frederick Smyth, produced the Fleet sewer engraving, his signature seen in the centre-bottom of the image (**4.2**);[24] while a number of anonymous engravers produced the 1859 image, each working on different parts of a larger woodblock divided up and then reassembled. The latter form of production mirrored that actually depicted in **4.3(a)**: the many different workers perform their distinct tasks to produce a coherent finished product – that is, the new sewers. Such a correspondence was one of the reasons why the editors of the *Illustrated London News* consistently stressed the appropriateness of wood engraving to depict precisely this kind of technical subject matter. The reference in the article to the engraving being based on a photograph also stresses its documentary quality, even if this is offset by the compositional changes made. Furthermore, the text of the article, seen in the top-half of the page, corresponds very closely to the engraving in the way it describes the scene depicted, educating the reader as to its context in the drainage system as a whole. Once again then, if this image is dramatic and sublime, it is also an attempt to document and educate the newspaper's readers.

Sublime Workers

The active workers depicted in the 1859 engraving (**4.3: a**), in contrast to the passive figures seen in the photograph (**4.3: b**), not only represent the technical processes involved in constructing the sewers, but also reflect the newspaper's attitude towards this particular class of labourer. Photography, in the 1860s, had not yet developed as a 'snapshot' art, being still reliant upon a long exposure time and cumbersome equipment.[25] Figures in photographs from the 1860s are largely static, gazing at the camera, or ghostlike and blurred if caught in motion. However, in the 1859 engraving, active workers form an important aspect of the scene. In its articles on the construction of the main drainage system, the *Illustrated London News* consistently commented on the large number of workers involved in construction – upwards of 2,000 on the northern high-level sewer alone.[26] In this engraving, the supervisors of the works are differentiated, in their top hats and black clothing, from the labourers, or 'navvies', seen atop the sewers, at the edges of the scene both looking at the works and towards the viewer, and in the foreground directly engaged upon the works. The two bricklayers in the centre foreground are the focus of the whole composition: a prominent representation of work that 'celebrate[s] a new form of heroic, manly labour' – that of metropolitan improvement.[27] If the forms of construction are configured as sublime, the navvies here both create and manage these sublime forms; this is the source of their perceived nobility. In short, alongside their educative role, the navvies also form a key part of the rhetoric of improvement that defined the attitude of the *Illustrated London News* towards the new sewers.

This 'ennoblement' of the navvy formed an important aspect of middle-class Victorian attitudes towards metropolitan improvements.[28] The construction of Bazalgette's sewers in the 1860s occurred alongside other vast building projects in London and the navvies, numbering tens of thousands, would have been a ubiquitous presence in the city during this time, especially in the eyes of middle-class observers. For some of these, progress became synonymous with the workers who ushered it in. Ford Madox Brown's (1821-93) painting *Work* (1852-65; **4.5**), with its accompanying text by the artist, is perhaps the most overt representation of the 'noble' navvy identified with progress. Tim Barringer's recent wide-ranging analysis of *Work* had shed light on a variety of possible interpretations of the navvies seen excavating a trench in the foreground of the painting. For Barringer, the navvy in *Work* represents at once the modern equivalent of the heroic male body seen in classical history painting; a call for the religious transformation of man through hard work; a challenge to the existing political order; and a beautification of the masculine.[29] However, despite the comprehensiveness of his analysis, Barringer, like most other commentators on the painting, has made one incorrect observation: the navvies in *Work* are not laying a water pipe, as has been assumed, but a sewer – both the significant depth of the trench and also the use of brick confirm this.[30] Consequently, commentators have failed to link the exhibiting of *Work* in 1865 with the enormous public interest in the main drainage system, which was partly completed in that year. So when, in 1865, the *Illustrated London News* praised Brown's painting as representing 'the principal hero: that potent agent in the work of British civilisation', it was not only identifying the navvies central role in the national mission of industrialisation and progress but also, specifically, their vital role in the construction of the main drainage system.[31] Indeed, compared to the colossal scale of the main drainage project, already discussed, the excavation undertaken in *Work* – a local sewer in Hampstead in 1852 – must have seemed to the *Illustrated London News* a very minor form of improvement to that which it was championing when the painting was exhibited in 1865.

In fact, the noble navvy was an emblematic figure in wood engravings in the *Illustrated London News* long before Brown began his painting in 1852. As early as 1843, the newspaper had presented the navvy in equally glowing terms. After the opening of the Thames Tunnel in March 1843, the newspaper depicted a navvy in repose (**4.6**), which formed the header for a long description of the history of the Tunnel.[32] This figure, with his rippling muscles and graceful reclining pose holding a medallion portrait of the engineer of the Tunnel, Sir Marc Brunel, is reminiscent of an ancient god or athlete – his shovel representing the 'noble' emblem of his work. However, as the newspaper made clear, this image represents one of the ordinary miners 'in his working dress, the long cap hanging down the back to protect him from the dripping of water'.[33] Here, as in *Work*, the status of the navvy is raised above the typical middle-class views of these workers as drunken, violent and sexually promiscuous.[34] The *Illustrated London News* simultaneously brings down-to-earth the heroic, god-like figure found in classical sculpture or history painting and transforms the status of the navvy to something far nobler than expected. This resting figure has only been beatified by his work – the effects clearly evident in the muscles of his body and the serene expression on his face.

Such depictions are characteristic of the *Illustrated London News*'s coverage of the new engineering projects of the 1860s; in addition to the main drainage system, these included the Metropolitan Underground Railway, the Thames Embankment, and several new overground railway lines and bridges. But here, it was the working, not resting, navvy that dominated its illustrations. In particular, during the construction of the Tower Subway in 1869 – described as 'the new Thames Tunnel' – both the *Illustrated London News* and its umbrella publication, the *Illustrated Times*, published several wood engravings of the work in progress.[35] One notable example, a striking front-page image in the *Illustrated Times* on 18 September (**4.7**), shows five navvies advancing the iron shield employed to form the Tower Subway – a tunnel 400 yards beneath the Thames from Tower Hill to Pickleherring Street in Southwark. The navvies in this image are immediately reminiscent, in both composition and pose, of those seen in Brown's *Work*. Like Brown's heroic male figures 'frozen in action', they clearly exemplify the ambition of the illustrated newspapers: to produce a form of everyday history painting; in effect, to heroicise the modern and those who implemented it.

Returning to the workers seen in **4.3(a)**, we can now more clearly appreciate their place within the particular iconography of the navvy developed by the *Illustrated London News* from the early-1840s onwards. If the photograph (**4.3: b**) shows the navvies blankly gazing, passive or slumped in exhaustion, the engraving rather depicts them frozen in action, ennobled by their 'heroic' task. And yet, these figures, like those in *Work*, seem to be deliberately set against others: the working figures – the bricklayers in the foreground and atop the arches of the sewers – contrast sharply with the supervisors and other navvies passively standing behind them. In fact, what we see in this engraving are some figures *looking at* the bricklayers – a situation reminiscent of that seen in *Work*, where two intellectuals watch, and ponder on, the navvies digging the sewer

4.5 Detail of Ford Madox Brown's painting *Work* (1852-65) showing workmen – or navvies – digging a trench and constructing a new sewer in Hampstead.

trench. If, as Barringer suggests, the working navvies in *Work* represent a direct challenge to the supposed social superiority of both intellectuals and the idle upper classes, what might we say about the relationship of the figures in the engraving?[36] The engraving focuses attention both inside (from the supervisors and excavators) and outside the image (from the supposed 'respectable' middle-class readers of the newspaper) at the particular work of the bricklayers. By doing so, the newspaper highlights this 'constructive' work: that is, the process by which the new sewers were made visible. Neither the excavators or supervisors could provide this visual sense of construction-in-progress; excavators were unskilled labourers who performed a more destructive kind of work, while the supervisors merely observed those at work – their passive forms almost identical in both photograph and engraving. For the *Illustrated London News*, bricklaying represented a powerful, and immediately visible, way of depicting what it regarded as an essential and noble form of improvement. It also allowed the newspaper to assimilate sublimity with humanity; if the new forms of construction threatened to overwhelm the city in their magnitude, the army of bricklayers seemed to represent the orderly control and mastery of these mythic forces. This idealisation of this particular 'constructive' navvy forms an important element in the articulation of an ideology of improvement, which characterised the attitude of the *Illustrated London News* towards technological change in London.

The Experiential Sublime

On 8 and 9 October 1861, London's press were given an opportunity to experience the construction project at first hand, through tours organised by Bazalgette and the Metropolitan Board of Works. These tours included inspections of the works at Wick Lane, the northern outfall sewer, the outfalls at Barking and Crossness, the Woolwich tunnel, and finally the works at the Deptford pumping station. The visits formed a focal point for press interest in the works and most of London's non-illustrated daily and weekly newspapers published voluminous articles on the days following the tours.

The language of these articles ranges from the explanatory to the poetic. Such a variety of expression reflected the need for these newspapers not only to explain Bazalgette's scheme to their readers but also to embellish such

4.6 One of Marc Brunel's workers on the Thames Tunnel as depicted by the *Illustrated London News* on 26 March 1843, p. 226, and holding the emblem of the engineer's company in the form of a medallion.

4.7 Five workers advance the metal shield in the construction of the Tower Subway under the river Thames, *Illustrated Times*, 18 September 1869, frontispiece.

explanations with appealing language that perhaps had a similar 'illustrative' function to the engravings in the *Illustrated London News*. All praise the works in glowing terms, stressing their sheer scale and magnificence of purpose. Some of the articles even ascribe a mythic status to the drainage works: according to the *Weekly Times*, the new sewers rival, even exceed, the 'seven wonders of the ancient world'[37] or 'the aqueducts and *cloacae* of ancient Rome'.[38] The Metropolitan Board of Works is described as the new 'Hercules;'[39] Bazalgette, as its spokesman, the new 'Cicero'.[40] Such language mirrors the rhetorical tone seen in the *Illustrated London News*, but here it is the experience of the vast sites of construction that provides the impetus for such assertions: the northern outfall sewer is wide enough 'to admit a horse and cart, and you could ride up [it] on horseback for miles', its straightness like the undeviating line of a Roman road.[41] The descent into the Woolwich tunnel, running for a mile beneath the town, was accompanied by the laying of the final brick of the tunnel with a silver trowel by John Thwaites, the chairman of the Board of Works.[42] Such ceremonial additions lent the visit an elevated, symbolic status, which then pervaded the subsequent press accounts.

However, if the new sewers are described as a modern 'wonder of the world', they are, in the same accounts, presented in a more disturbing light: they are also described as 'rapacious',[43] devouring millions of bricks, and stretching 'like a girdle round London'.[44] The final destination of the visit, the works at Deptford pumping station, included the sight of 'divers ... plying hammer and pick forty feet under water'.[45] The *Daily News* comments that 'the miles upon miles of brickwork on the acres of planned land, the steam-engines, the temporary timber structures, the wagons, the carts, and the thousands of busy workmen' produced 'a succession of effects quite bewildering to the instructed eye'[46] with the workers 'swarming like bees' and making up a potent element in this scene of unintelligible confusion.[47] If this confusion is an imposing spectacle, it is also bewildering and disturbing to the senses, serving to highlight other negative aspects of the experience of the sublime. In these articles, the close-hand experience of the works and the workers generates complex and often-contradictory emotions, ranging from celebration to anxiety, and expressed in language that mirrors the shifting qualities of the sublime as a first-hand experience.

On 30 November 1861, the *Illustrated London News* published a special supplement illustrated by eleven engravings detailing the works visited – by far the most comprehensive series of illustrations of the main drainage works. The text of the accompanying article is prosaic, celebratory and explanatory. The engravings, by contrast, articulate the same sense of ambivalence seen in the non-illustrated press accounts. **4.8** shows two engravings included in the supplement, depicting views inside the sewer tunnels of the southern high-level sewer at Peckham in south London. The right-hand image represents two workers at rest in a completed section of tunnel, while in the engraving on the left a single worker excavates with a pick while another watches, seated in the foreground. The differences between the two images are startling: in the right-hand image, the workers are symmetrically framed in the massive circle of the sewer's cross-sectional form, the patch of light falling from above accentuating their resting forms; the image on the left is uniformly dark, the light seen behind the digging figure only serving to silhouette his dark form. The space inside this sewer is claustrophobic, emphasised by the repeating octagonal forms of the timber support and the struts converging in front of the digging figure. The formless mound of excavated material in the foreground, in contrast to the ordered clarity of the image on the right, gives the impression of a dirty and disordered working environment. In short, the subterranean imagery here is more akin to a vision of a demonic underworld than an ordered or noble space.

Such a dramatic contrast in the representation of underground spaces can also be detected in earlier accounts of visits to subterranean excavations. In 1827, the English actress Fanny Kemble (1809-93) described, in a letter to a friend, a visit she made to the Thames Tunnel, then being excavated by navvies.[48] She begins by describing the tunnel as a 'vast, illuminated, silent

4.8 Two contrasting views of workers constructing the southern high-level sewer in Peckham in the *Illustrated London News*, supplement, 30 November 1861, p. 554.

fairyland' and the excavators as 'beautiful, wise, working creatures'. In this space Kemble experiences 'amazement and delight'.[49] However, as she progresses into the darker recesses of the tunnel her reaction changes. The decisive moment comes when she views the navvies in the depths of the darkness, in the act of excavation. She describes them as 'all begrimed ... some standing in black water up to their knees, others laboriously shovelling the black earth in their cages ... with the red, murky light of links and lanterns flashing and flickering about them'. Such a scene, for Kemble, although a 'striking picture', also conjures up an image of 'the beautiful road to Hades'.[50] Crucially, it is the sight of the navvies, working in brutal and infernal conditions, that gives rise to this reaction. If Kemble is describing a sublime experience, it is an experience characterised by both wonder and terror – the movement from one state to another being related to the changing perspective of both the viewer and the spectacle viewed.

Returning to the engravings (**4.8**), in the light of Kemble's account, some further comments can be made on the differences between the two images. In

the right-hand image the workers are depicted at rest, their bodies showing no signs of any adverse effects of their brutalising work; in the image on the left is shown something of the actual conditions and experience of a more destructive kind of work than the ordered bricklaying seen in the earlier engraving (**4.3: a**) – in this case the digging of an enormous trench by hand through both soil and rock. Like the compositional arrangement of **4.3(a)**, we witness a passive figure looking at one who is working. Yet, what the figure and reader are looking at is very different from that seen in the earlier engraving; the single figure working in the darkness seems out of proportion to the scale of the work undertaken and his work is rendered illegible by the uniform darkness of the engraving. With its focus on atmosphere rather than clarity, this image reflects a different view of 'sublime' work similar to that experienced by Kemble: a move from a sense of ordered serenity, seen on the right, to one of infernal confusion and brutality, seen on the left.

The foregoing discussion of the dramatic qualities of these two engravings has not addressed the question of their possible function as documentary images – an equally important role of wood engravings according to the *Illustrated London News*. In relation to this particular edition of the newspaper, the two images are both part of a single page, with its two additional engravings (**4.9**), and seven more engravings and descriptive text arranged over three more pages.[51] In this context the two engravings of the sewer tunnels, as well as being dramatic in their own right, are constituent parts of a wider series of images that depict all the sites visited by the press; as such they are part of a comprehensive attempt to educate the viewer as to the various elements of this spectacle of construction. Both images can also be read as 'before' and 'after' illustrations of the same scene: one showing the driving of the tunnel; the other, the finished product, with the two workers, presumably the same in both images, serving to educate the viewer as to the processes involved in the construction of an underground sewer tunnel. Nevertheless, compared with the 1859 engraving (**4.4**), the 1861 images demonstrate an increasing dominance of the visual. Furthermore, the engraving technique of the two images of the sewer tunnels (**4.8**) is very different from the 1859 image with its precise delineation of work and workers alike: rather, in these images heavy black lines obscure any sense of illustrative clarity; atmosphere seems to be more important than technical exposition. All explanatory text in relation to the engravings, bar small captions, is removed to separate pages, but even here there is very little explanation as to the content of the images. Like the other images discussed, there is here a relationship between the dramatic and the prosaic, but one in which the balance between the two has significantly shifted. These two engravings – in terms of their iconography, technique and context – stress the dramatic over the descriptive and articulate a form of subterranean sensationalism that draws, like Fanny Kemble's account, on different and darker aspects of the sublime that relates to a first-hand experience rather than an ideology or rhetoric of improvement.

The Sensational Sublime

The engravings examined so far suggest a shifting relationship between their roles as forms of technical exposition and vehicles of sublimity (whether in a rhetorical or experiential context). Representations of accidents that occurred during the construction of the main drainage brought out the dramatic possibilities of wood engraving most strongly, by means of a sensationalist and titillating iconography of destruction. Accidents provided a focus not only for fears about the potential destructive power of technology, but also for the enjoyment of this destruction (and the fear it provoked) from a safe distance.

One of the first serious accidents occurred on 28 May 1862, when an explosion at Shoreditch – caused by a ruptured gas main next to the excavation for the southern high-level sewer – resulted in localised destruction and the death of a passer-by. It is perhaps significant that the *Illustrated London News* did not report on this event. When the newspaper did report on industrial accidents its tone was, on the whole, restrained and sombre, the accompanying illustrations often giving only a limited sense of the destruction caused.[52] No doubt this restraint was meant to reassure its middle-class readers of the benefits of new technology, in line with the ideological bias of the newspaper. The cheaper penny illustrated press, which appealed to a much less 'respectable' audience, had no such reservations: both the *Penny Illustrated Paper* and the *Illustrated Weekly News* often published graphic images of violence on their front pages; bodies were depicted blown into the air or mangled in the remains of explosions.[53] Such sensationalism characterised their reportage of the Shoreditch accident. In almost identical accounts, both newspapers describe the accident in lurid detail: the *Penny Illustrated Paper* states that when the gas main was ruptured:

> ... the gas blew out with a loud noise, and flew through the open ground along the sewer until it reached the furnace of the engine. A number of men in ... the cutting were instantly prostrated. A woman named Hannah Smith, who was passing along the pavement, was knocked down and her clothing set on fire.[54]

After these events, one of the houses flanking the open cutting:

> ... was blown to the ground as if struck by a shell from a mortar, and the fragments falling upon the poor woman, she was held fast in the midst of the flaming mass until extricated by the firemen.[55]

Nine more houses 'had their fronts blown out' and the unfortunate Hannah Smith later died as a result of her injuries.

These accounts stress the devastation, both to person and property, that occurs when technology fails. The gas that escaped, according to the *Illustrated Weekly News*, 'rushed out with a noise resembling a perfect hurricane' before which men were struck down 'instantly'.[56] As for Hannah Smith, she is an unwitting victim of this technological failure: a passer-by who is horribly mutilated by both the gas and the falling debris of the destroyed houses. Here the navvies, rather than being masters of the works they have created, are instead victims of its destructive power, unleashed without warning. Such a power corresponds to that hinted at in the articles and illustrations considered previously, but here the sense of terror is unmediated by admiration or awe and is expressed in lurid language that draws on sensationalist aspects of the sublime, as a means of thrilling their working-class readers.

On 18 June 1862, another more spectacular accident attracted the attention of the majority of London's illustrated newspapers: the bursting of the Fleet sewer. It was also, significantly, given uncharacteristically detailed coverage by the *Illustrated London News*. In Victoria Street, Clerkenwell, the Metropolitan Underground Railway was being constructed adjacent to the Fleet sewer, which was then being excavated in order to direct its flow into the completed northern middle-level sewer. After heavy rain on 18 June, the inundation in the sewer caused its walls to breach, flooding the railway trench and eventually the entire surrounding district. The resulting spectacle was described in vivid detail in press accounts. According to the *Illustrated Times*, 'the scene of the accident, like the scenes of all great accidents, is rather impressive'.[57] The *Penny Illustrated Paper* reports that, when the water had initially entered the railway trench, 'men were lowered into the tunnel in baskets' in order to break holes in the cutting to allow the water to escape.[58] However, as the *Illustrated London News* states, catastrophe was to follow when 'the cracking and heaving mass [of the trench]

LONDON MAIN DRAINAGE.

WORKS AT BARKING CREEK OUTFALL.

PUMPING-STATION AT DEPTFORD CREEK.

DRIVING A TUNNEL AT PECKHAM.

BOTTOM OF A SHAFT IN THE SOUTHERN HIGH-LEVEL SEWER AT PECKHAM.

4.9 Page layout from the *Illustrated London News*, supplement, 30 November 1861, p. 554, showing a range of engravings depicting the construction of the main drainage system.

… rose bodily from its foundations' and was lifted 'into one heap of ruins'. By six o'clock in the evening 'thousands of people had collected round the spot' in order to 'enjoy the full view of the ruin'.[59] This view is described in lurid detail: '[t]he fallen roadway, with the bent lamp-post and pavement, looks as if it had been sucked down by a whirlpool;'[60] gas and water mains lay strewn about while the flood waters, 'inundate[d] the cellars of the houses … appearing to have uncontrolled power over the whole portion of that low lying district'.[61] The walls of the railway cutting were 'snapped off from their foundations, like pieces of broken matchwood', while 'the black hole of the Fleet sewer, like a broken artery, pour[ed] out a thick rapid stream which found its way out fiercely … into the railway cutting'. Most disturbing of all was the inundation of a paupers' burial ground: one of the tombs had a corner 'knocked off by the uplifting struts, and sufficient damage [was] done to excite the imagination of the curious'.[62]

These accounts, laden with vivid symbolism, focus on the destructive power of nature, rather than that of technology (seen in the earlier gas explosion at Shoreditch). Such power creates a dramatic spectacle that is not only enjoyed but also the result, according to the *Penny Illustrated Paper*, of 'fears … fatally verified'.[63] This sense of fear is heightened by the symbolism employed in the articles: the floodwaters form a 'whirlpool', the sewer is like 'a broken artery', the railway works are like 'broken matchwood'. The exposure of dead bodies gives a peculiar sense of morbid fascination to the scene – the unleashed powers of nature, mediated by technology, can even disturb the dead.

These descriptive accounts were given a visual equivalent in almost identical engravings published in the *Illustrated London News* and the *Illustrated Times*. The former published its large and dramatic image on the front page of the 28 June issue (**4.10**). Unlike most of the newspaper's representations of accidents, and in common with its more sensationalist and less 'respectable' rivals, the viewpoint of this engraving is not removed to a safe distance from the scene of destruction: on the contrary, the broken timbers and dark waters of the sewer lie very close to the observer; we, the viewers, are, according to the *Illustrated Times*, precariously situated at 'the edge of the planking which overlooks the deep cutting'.[64] A few small figures are seen in the middle distance trying to manage this scene of vast confusion, watched by the crowd of onlookers behind. Unlike the workers seen in the 1859 engraving (**4.3: a**), these figures are almost illegible; they seem to merge with the crowd seen in the background with no attempt to clarify the work they are performing. Instead, the destruction itself dominates the image. Depicted from left to right are: the 'black hole' of the Fleet sewer, the damaged walls of the Underground Railway, and the exposed gas mains, water pipes and buried lamppost. Closely matching the descriptive details of the accompanying article, the image situates the viewer in disturbingly close proximity to danger. If the scene of destruction induced an experience of the sublime, it is one that oscillates between enjoyment and fear: between the safe viewpoint of the crowd and our more dangerous position as viewers, close-up to the scene of destruction.

This engraving also suggests a further sense of the dominance of the visual,

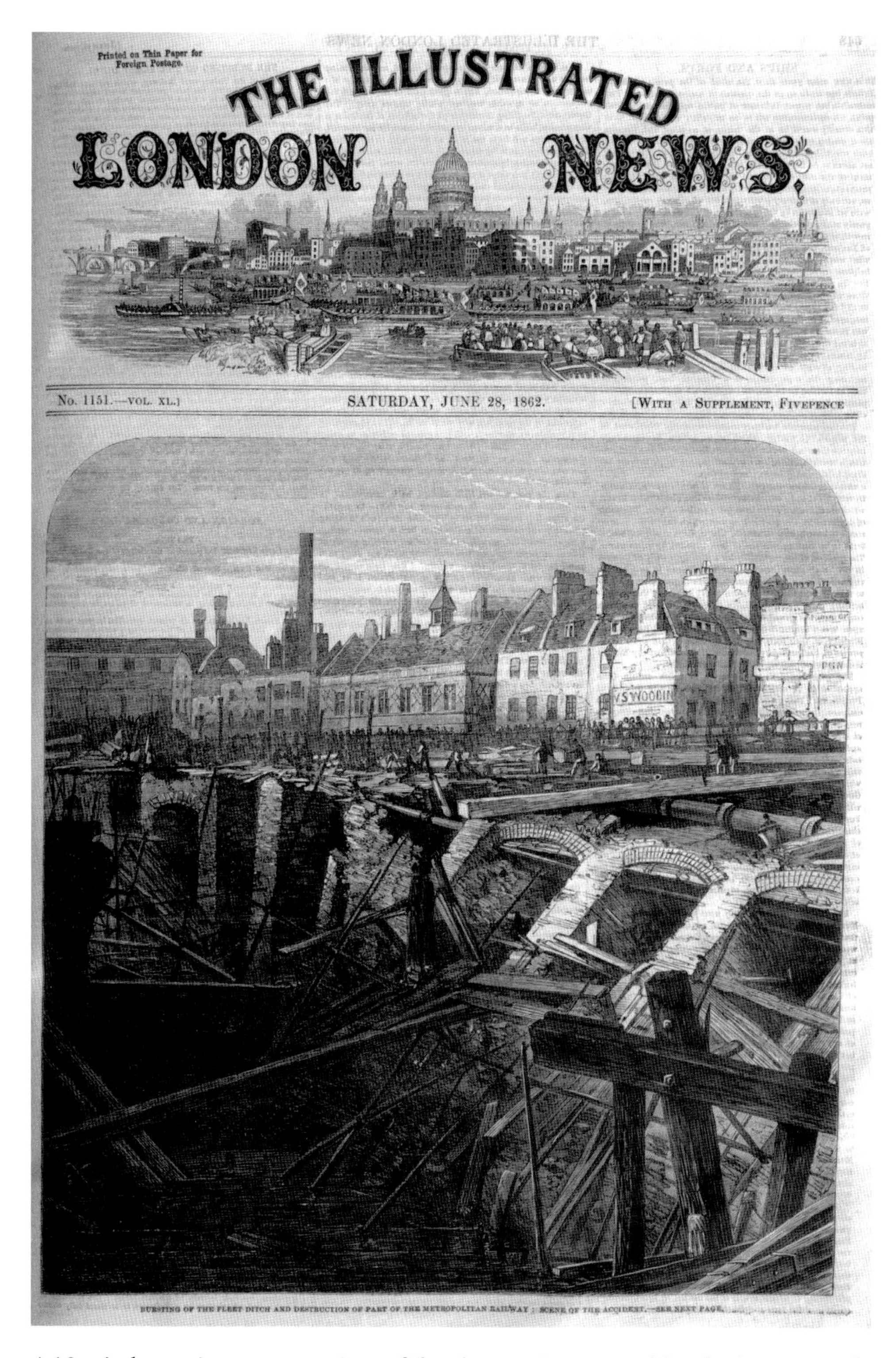

Printed on Thin Paper for Foreign Postage.

THE ILLUSTRATED LONDON NEWS,

No. 1151.—VOL. XL.] SATURDAY, JUNE 28, 1862. [WITH A SUPPLEMENT, FIVEPENCE

BURSTING OF THE FLEET DITCH AND DESTRUCTION OF PART OF THE METROPOLITAN RAILWAY: SCENE OF THE ACCIDENT.—SEE NEXT PAGE.

4.10 A dramatic representation of the destruction caused by the bursting of the Fleet ditch on the front page of the *Illustrated London News*, 28 June 1862.

compared with the other engravings examined in this chapter, and consequently of the dramatic over the demonstrative. As a front-page, stand-alone image, it dominates not only in terms of its large size, but also as the defining image of this particular edition of the newspaper: it would have been the first thing any potential reader would have seen. Employing a similar technique to that seen in the 1861 engravings (**4.8**), the image is at once obscure and dark but also, in this case, disturbingly detailed. Unlike the 1861 images, with their lack of explanatory text, the iconography of this engraving corresponds closely with its accompanying article on the next page, but without any sense of the educational value of the earlier engravings. On the contrary, this close interrelationship of text and image rather enhances the sensationalist power of the image – the attention to detail serving to titillate rather than to educate. Here the *Illustrated London News* perhaps vies with its less 'respectable' competitors for a wider audience by adopting their own sensationalist approach.[65]

The accidents examined above represent but one instance of a wide coverage of technological failure in the illustrated press throughout the 1860s. The building of the new sewers and railways in London in the 1860s resulted in the literal tearing apart of the urban fabric and its existing communities to make way for their 'improving' work.[66] If the new sewers were seen as noble works, symbolising and ushering in great progress, they were also subject to the fears that govern attitudes towards new technology. Accidents provided the evidence and outlet for these fears. Press responses to accidents such as the bursting of the Fleet sewer, whether in the cheaper, less 'respectable' newspapers or the more high-minded and expensive *Illustrated London News*, were couched in a sensationalist visual and verbal language of sublimity that was also an effective way of selling this imagery to an audience eager for such excitement. Yet this exciting spectacle of construction, whether viewed as constructive or destructive, was relatively short-lived; once building had finished – in 1865 on the south side of the Thames; 1868 on the north side – these vast tunnels became forever invisible, buried beneath the city. However, an important part of the main drainage system did remain visible, that is, the pumping stations. Their vital role in the promotion of the project forms the focus for the rest of this book.

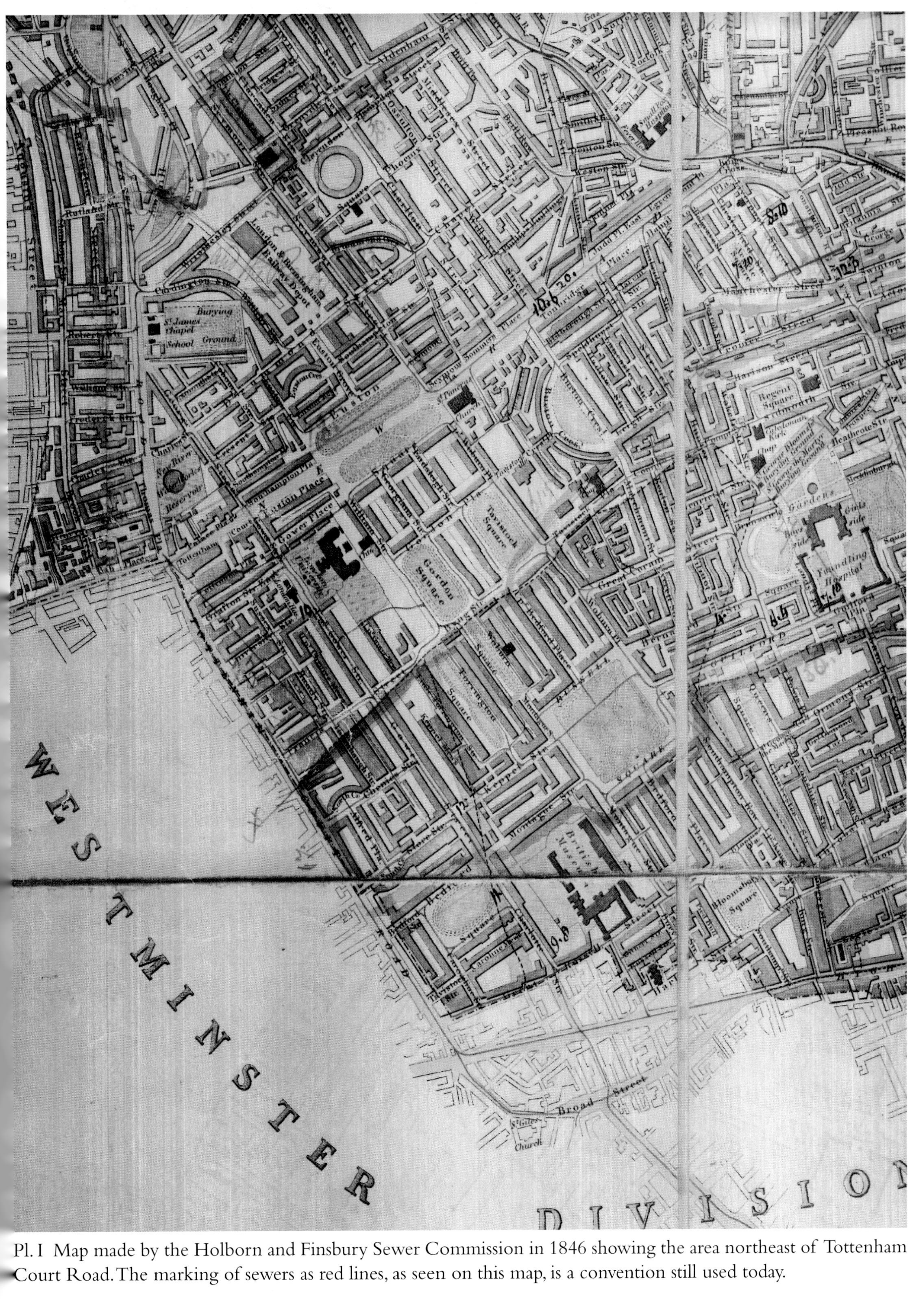

Pl. I Map made by the Holborn and Finsbury Sewer Commission in 1846 showing the area northeast of Tottenham Court Road. The marking of sewers as red lines, as seen on this map, is a convention still used today.

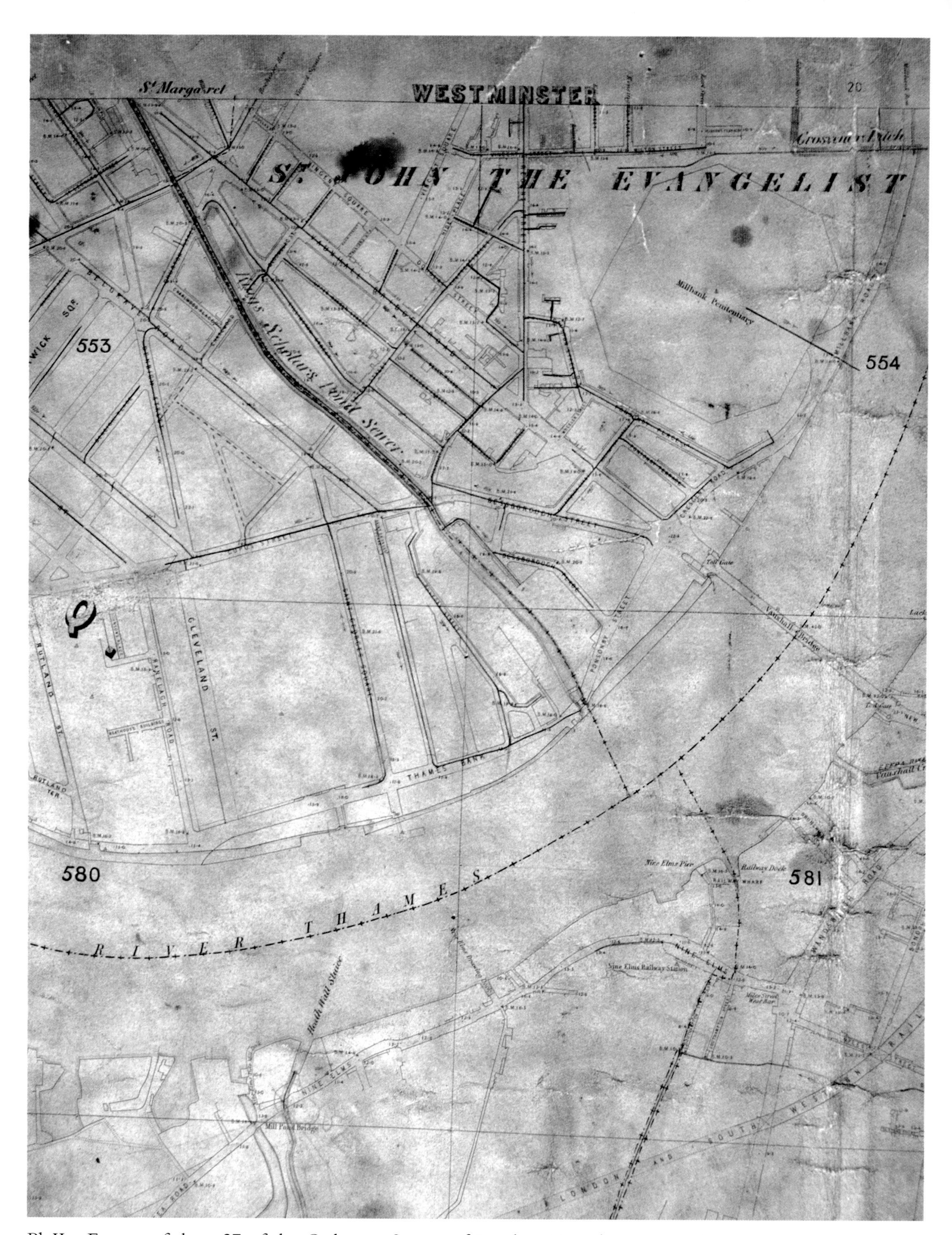

Pl. II Extract of sheet 27 of the Ordnance Survey of London 12-inch to one-mile skeleton plan with sewer overlays, which was drawn up around 1850 by the Metropolitan Commission of Sewers after they abandoned production of the ten-feet to one-mile maps.

Pl. III Frank Forster's proposed plan for the drainage of north London included in his report presented to the Metropolitan Commission of Sewers on 30 January 1851.

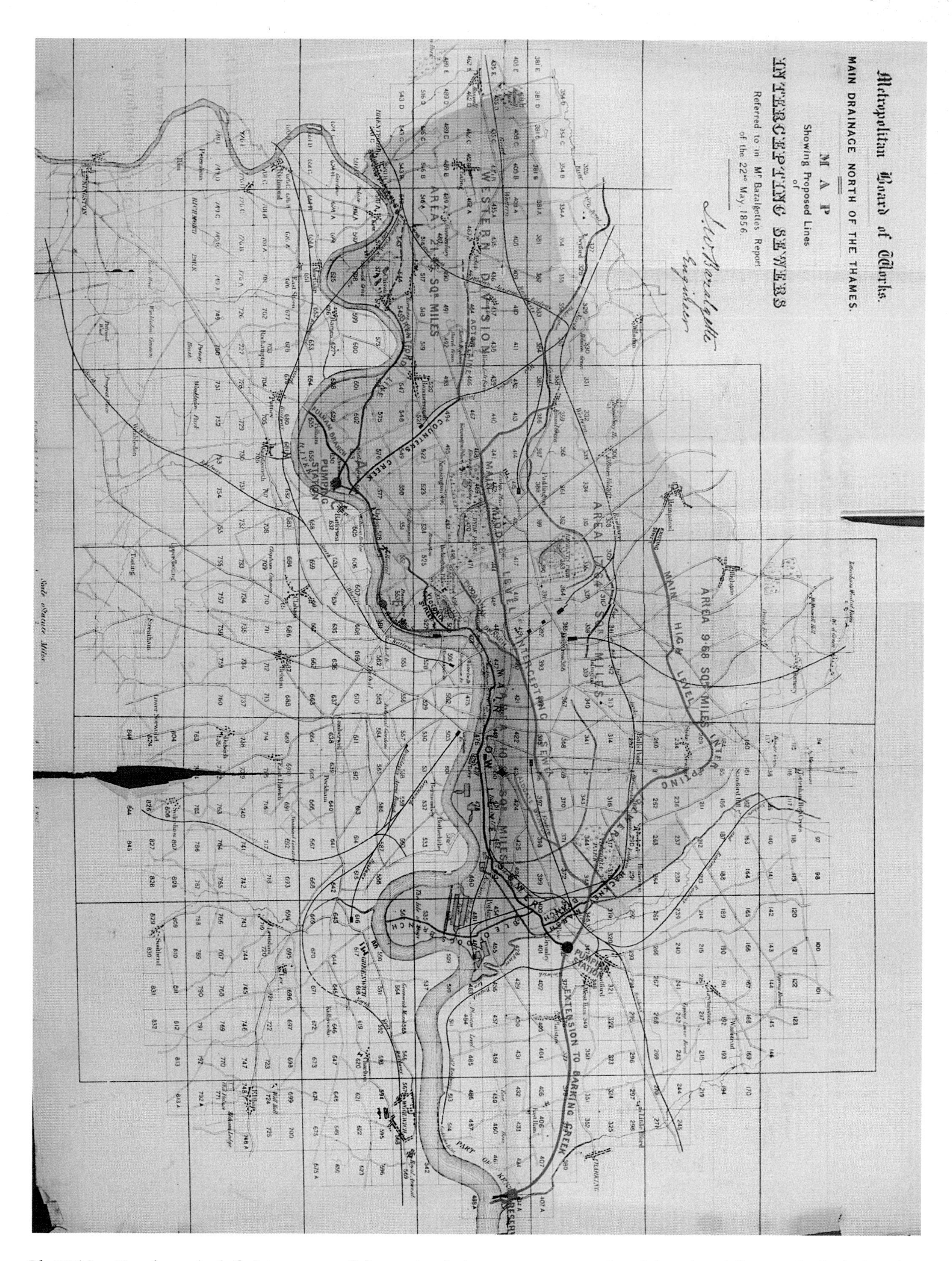

Pl. IV(a) Bazalgette's definitive map of the main drainage system north of the river Thames included in his report presented to the Metropolitan Board of Works on 22 May 1856.

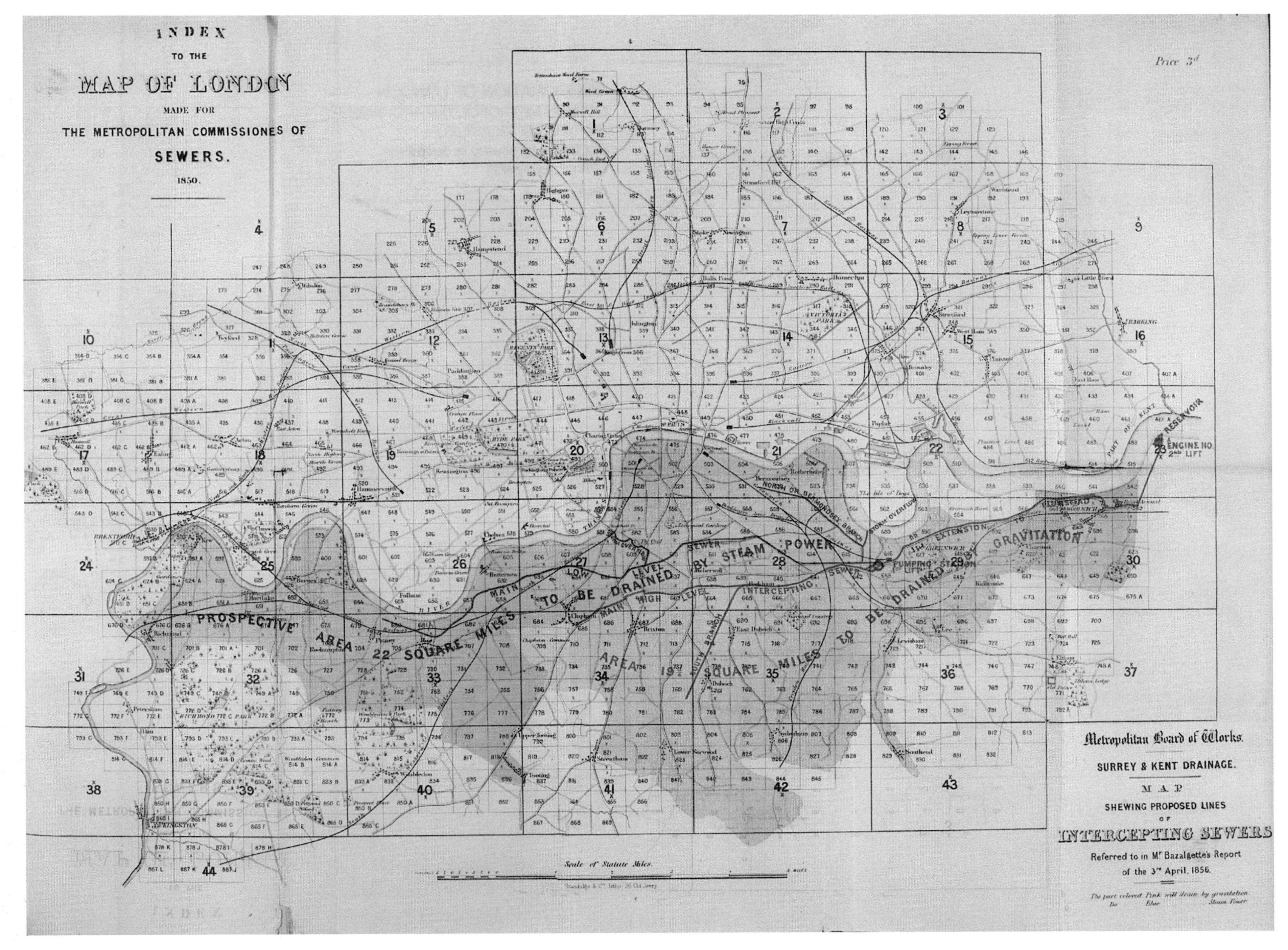

Pl. IV(b) Bazalgette's corresponding map of the southern drainage presented on 3 April 1855.

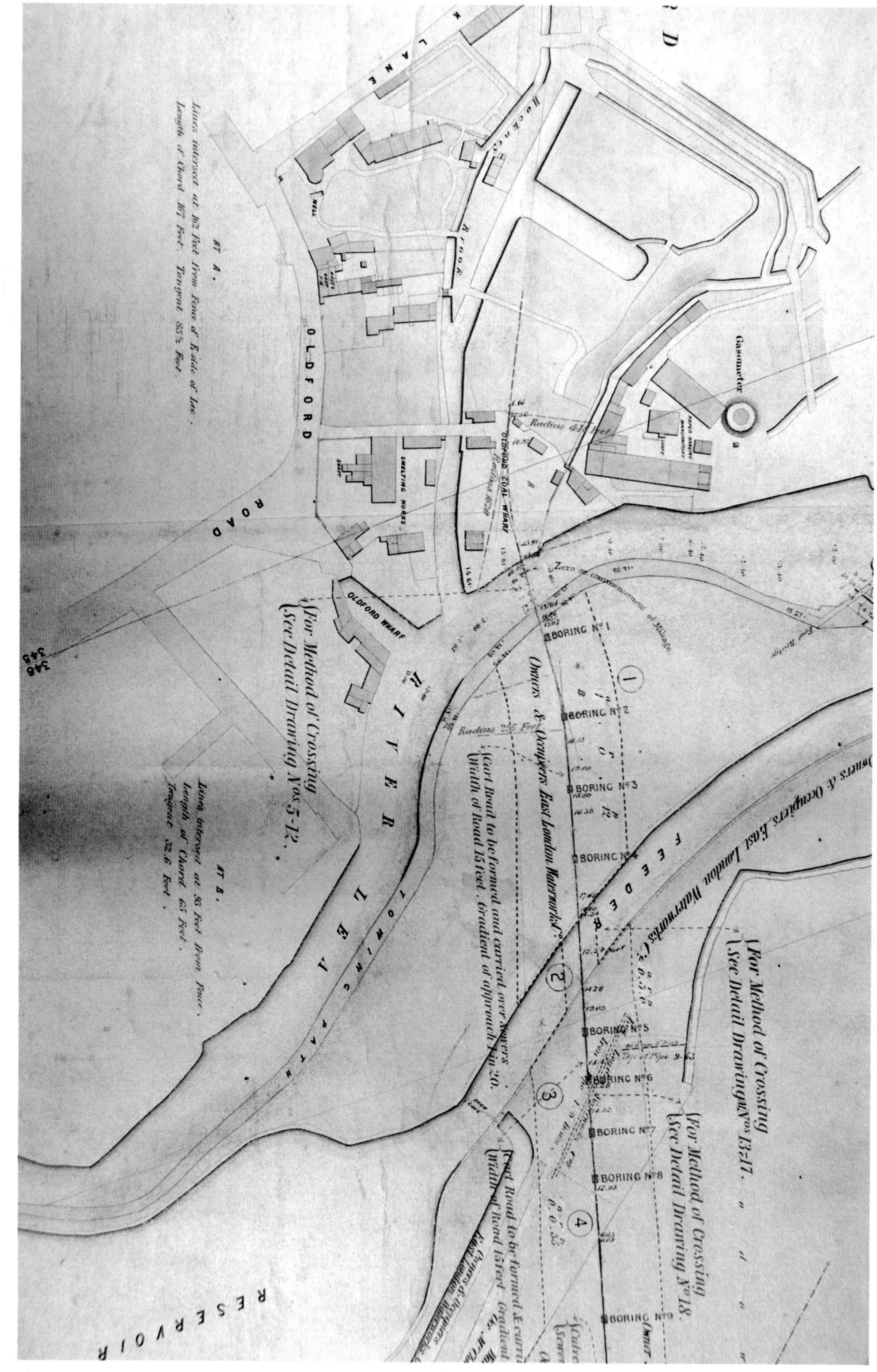

Pl.V Drawing no. 3 in the northern outfall sewer contract showing a one-hundredth extract of the general plan which in total measured around ten metres in length and was hand-coloured throughout.

Pl.VI(a) The west bay of the Abbey Mills pumping station, showing rich use of coloured tiles in the cornice, and bricks and stone in the windows and porch.

Pl. VI(b) View of the train shed of London Bridge railway station along St Thomas's Street, built in 1866 and designed by Charles Driver.

Pl.VI(c) Driver's Battersea Park railway station, built in 1866 and mixing Italian (round-arched windows) and French (segmented windows) styles in its three storeys.

Pl. VII(a) One of the four identical porches of the Abbey Mills pumping station, designed by Charles Driver, and showing his unique quadripartite hood.

Pl.VII(b) Driver's similar porch at Leatherhead railway station.

Pl. VIII(a) Encaustic tiles in the recesses beneath the first floor windows of Abbey Mills, just one of numerous designs used in the exterior of the building.

Pl. VIII(b) Encaustic floor tiles in the chancel floor of St Mary's Church, Warkworth, Northamptonshire, installed in 1868 just after the completion of Abbey Mills.

Pl. VIII(c) One of Owen Jones's designs for medieval tiles in *The Grammar Of Ornament* (1856), plate LXX, no. 12.

Pl. VIII(d) A more elaborate design from the same plate in *The Grammar of Ornament*.

Section III: Architecture

5.1 The exterior (upper engraving) and interior (lower engraving) of the Abbey Mills pumping station as depicted by the *Illustrated London News* on 15 August 1868, p. 161.

5
Engineering and Art

Architecture played a vital role in both the function and the public promotion of the main drainage system, in particular the pumping stations at Crossness (1862-65) and Abbey Mills (1865-68), which were key sites for important ceremonies held to mark the completion of the system. This chapter will concentrate on the architectural significance of Abbey Mills and will bring to light, for the first time, the role of its architect, Charles Driver (1832-1900). The following and final chapter will shift the focus to responses to the pumping stations and will discuss the importance of Abbey Mills and Crossness as focal points for public awareness of the main drainage system. Both of these chapters will develop new ways of understanding these magnificent examples of Victorian industrial architecture.

Architectural historians have consistently commented on the flamboyant eclecticism of Abbey Mills (**5.1**); it has variously been described as 'vaguely Italian Gothic',[1] 'Moresque',[2] 'Slavic',[3] 'Byzantine almost Russian',[4] 'Renaissance/ Second Empire style with slightly Eastern features',[5] or resembling a 'Greek Orthodox cathedral'.[6] Up until now, these historians have ascribed the design of Abbey Mills to either Bazalgette or his assistant engineer Edmund Cooper. Such a conclusion has generally resulted in the assessment of Abbey Mills as an uneducated, whimsical – even vulgar – architectural statement, characterised by an apparent arbitrary eclecticism as suggested by the descriptions above. These assessments have been partly redressed by the much-delayed attribution of the design to Charles Driver in the recently published revised edition of Nikolaus Pevsner's *Buildings of England* series for east London. However, there is no indication given as to why Driver has now been recognised as the architect and the meaning of what it, referring to the interior ironwork, describes as his 'outrageously hybrid art'.[7] So, who was Charles Driver and why is he only now recognised as the architect of Abbey Mills?

An Unknown Victorian Architect?

The death of Charles Henry Driver on 27 October 1900 was noted in the architectural press in the form of several extensive obituaries.[8] Celebrated as 'a man of very versatile talents'[9] and 'an authority on ornamental cast ironwork',[10] Driver was recognised for both his architectural work and also his engineering knowledge, the latter confirmed by his election to the Institution of Civil Engineers as an associate member on 12 February 1900.[11] Since his election to the Royal Institute of British Architects (R.I.B.A.) – as an associate in 1867 and as a fellow in 1872 – Driver had built up a significant architectural practice, based at offices in Parliament Street (1867-68) and Victoria Street (from 1872) in Westminster, at that time the heartland of London's architectural and engineering community.[12]

Most of Driver's work as an architect was characterised by a close collaboration with engineers. Such projects included numerous railway stations and bridges in England and later in Brazil, the largest of which is the celebrated 'Station of Light' in Sao Paulo (1897-1900); piers at Llandudno (1878), Nice, and Southend-on-Sea (1887-90); the aquarium and orangery at the Crystal Palace in Sydenham (1869-73; demolished);[13] and the cast-iron central market in Chile's capital Santiago (1868-70).[14] Among his works as sole architect are a public hall (1871, now the old fire station) and many shops and houses in Dorking;[15] the Horton Infirmary in Banbury (1869-72)[16] and the restoration and part-rebuilding of St Mary's Church (1868) in the nearby village of Warkworth;[17] the Ellesmere Memorial in Lancashire (1858);[18] and Mark Masons' hall in Great Queen Street, London (*c.*1870; demolished).[19] Before he established his own practice as an independent architect, Driver is recorded in documents in the Institution of Civil Engineers as having 'assisted … Joseph Bazalgette [from 1864 to 1866] … in preparing designs for the masonry and landing stages and the ornamental masonry for the Thames Embankment, and for the pumping stations at Abbey Mills and Crossness'.[20]

Despite such a comprehensive listing of his projects in his obituaries and nomination papers, much of Driver's work as an architect remains unacknowledged. Born on 23 March 1832, he began his career in 1850 as a draughtsman in the engineer's office of the Metropolitan Commission of Sewers; at this time many would-be architects began their careers in this way, as surveyors or draughtsmen working alongside aspiring engineers. It was here that Driver would have first met and worked alongside Bazalgette, who served on the Commission as an assistant surveyor from 1849 to 1850.[21] In 1852, Driver took up a similar position with the engineering partnership of Liddell and Gordon and it was here that he worked on his first architectural projects, assisting in the design of stations on the Leicester to Hitchin Railway, all of which were built in 1857 in an eclectic style mixing Norman and medieval Venetian elements and including elaborate cast-iron platform canopies.[22] From 1860 to 1863 he worked alongside Robert Jacomb Hood, engineer to the London, Brighton and South Coast Railway, designing stations on the Dorking to Leatherhead line,[23] the terminus stations at Portsmouth[24] and Tunbridge Wells,[25] and all of the stations on the South London Line from London Bridge to Victoria, all built in equally eclectic styles mixing Renaissance and medieval motifs in polychromatic brick and stone with a highly original treatment of cast iron.[26]

5.2 A photograph of the Deptford pumping station (1859-62) in the 1950s showing its twin engine-houses joined by a single-storey boiler house and prominent chimney in the background (now demolished).

Throughout Driver's career, his consummate skill as a draughtsman is evident in the many watercolour views he produced of his architectural projects. These include: an unexecuted design 'for an interior of a collegiate church', exhibited at the Royal Academy in 1855;[27] two views of the Crossness pumping station from 1864;[28] a watercolour showing the Horton Infirmary in 1872;[29] and a depiction of the magnificent central station in Vienna, designed in collaboration with the architect Joseph Fogerty in 1882.[30] Driver's versatility as both architect and artist did not go unrecognised in his lifetime, as indicated in his obituaries. Nevertheless, because he often worked in collaboration with engineers, his significance as a Victorian architect, in relation to some of his more famous contemporaries, remains difficult to assess; in most of his projects, including his work for Bazalgette on the Abbey Mills pumping station, the profile of the engineer took precedence.

The Main Drainage Pumping Stations

The four main drainage pumping stations – at Deptford (1859-62), Crossness, Abbey Mills, and Pimlico (known as the Western pumping station; 1870-74) – were, and still are, vital components of Bazalgette's sewerage system

5.3 The Crossness pumping station pictured in the *Builder*, 19 August 1865, p. 591, showing its two-storey engine house with Mansard roof and flamboyant chimney resembling an Italian campanile.

(**I.1**). Although the notion of pumping sewage at strategic locations in the city had characterised many earlier schemes for London's drainage, Bazalgette was the first to precisely define the location and projected specifications of the necessary pumping stations.[31] As he described, some areas of London, both north and south of the Thames, were below the high-water level of the river. Therefore, in order to remove sewage by mean of interception and to transport it downstream out of the city, at certain points the sewage would need to be pumped up to above the level of the Thames. Bazalgette used the points where his intercepting sewers joined as convenient locations for the necessary pumping to occur. At Deptford, the southern low- and high-level sewers joined; at Crossness the entire sewage of south London was pumped in order for it to drain by gravitation into the Thames. Similarly, on the north side, Abbey Mills marked the point where the northern low-level and outfall sewers met, with the Western pumping station acting as a subsidiary pumping point for the rapidly expanding low-lying western suburbs. Bazalgette's descriptions of the pumping stations, although they give precise specifications of their engineering function, provide no information as to architectural considerations – that is, in what style to build.

The four main drainage pumping stations were an entirely new building type, in that the main drainage system was the world's first urban sewerage scheme to use pumping power on such a scale. However, in terms of their function, the precedent was set by other types of pumping stations – for water and for use in mines.[32] The rapid development of steam engines after 1712 was in part propelled by their increasing employment as lifting devices in the mining industry. After James Watt invented his condenser and rotative engine in 1783, the resultant large-scale steam engines began to be housed in tailor-made structures known as engine-houses. The subsequent development of high-pressure Cornish engines after 1812 led to pumping engines of both enormous power and size, which began to be used in the water and, later, the sewage industries.[33] The immense size and weight of Cornish engines – some with up to 100-inch cylinders – meant that large buildings were needed to contain them. The engine-houses built for the London water companies at Kew Bridge (1838; **5.5**), Hampton (1855) and Stoke Newington (1856; **5.6**), in which the massive beam engines were built into the structure or 'shell' of the buildings, formed important precedents for Bazalgette's pumping stations in that their technological basis was the same.

However, in terms of their architectural style, Bazalgette's pumping stations represent somewhat of a departure from existing models. The first, at Deptford (**5.2**), is built to a simplified English Classical style; it consists of a pair of identical two-storey engine-houses joined by a single-storey boiler-house, decorated with simple panels and mouldings on the doors and windows, and a plain chimney-stack with a Tuscan base and cap. Crossness (**5.3**) departs radically from the restrained style adopted at Deptford: with its polychromatic brick and stonework, Flemish roof and ornate chimney, Crossness is boldly and lavishly treated. Abbey Mills (**5.1**) develops the design features seen at Crossness but with even more flamboyance and extravagance, while the Western pumping

5.4 The engine-house of the Western pumping station (1870-74) in Pimlico as it is today.

station (**5.4**) is built in a French château style with its simplified chimney and classical decoration; it represents another shift in style back to the less overt display seen at Deptford. Such eclecticism in part reflects the precedent set by water pumping stations, which are stylistically equally inconsistent. The engine-house at Kew Bridge (**5.5**), with its restrained classical style, resembles the later pumping station at Deptford. By contrast, the engine-house at Stoke Newington (**5.6**), designed by William Chadwell Mylne, and resembling a fantastical medieval castle, is perhaps more similar in spirit to Crossness or Abbey Mills.

If there is any consistency in the architectural treatment of the main drainage pumping stations, it is between Crossness and Abbey Mills, both lavish and flamboyant in their scale, decoration and high degree of architectural literacy. Location may have played a role in this: both Crossness and Abbey Mills, unlike Deptford and the Western, were, at the time of construction, situated in areas remote from the built-up parts of the city. If the restrained styles of Deptford and the Western fitted in with their surrounding buildings, then the isolated sites of Crossness and Abbey Mills gave free reign for a more striking and individual treatment. Furthermore, both Crossness and Abbey Mills – again unlike Deptford and the Western – were assigned the important role as key sites for public ceremonies associated with the formal opening of the drainage system south and north of the Thames – a 'public' role that certainly would have justified their lavish architectural treatment (see chapter 6).

Functional Space

The most comprehensive and influential description of the Abbey Mills pumping station is that by Bazalgette himself, who wrote an account of the building that was distributed to the visitors attending the opening ceremony on 31 July 1868.[34] Bazalgette's description is highly detailed, concentrating on the functional aspects of Abbey Mills, which he viewed as not only the most important of the four main drainage pumping stations but also 'the largest establishment of its kind in existence',[35] costing £269,620 (£212,000 for the buildings and £57,620 for the machinery).[36] Bazalgette's account describes the seven-acre site (**5.7**), which is divided up into two portions by the northern outfall sewer which passes diagonally across it on an embankment. Southeast of the embankment is the engine-house with its adjoining boiler-houses (one of which is extant), workshop and chimneys (demolished in 1940). Other buildings on the site included a coal-store, a wharf for landing coals and a house for the superintendent of the site (extant). Northeast of the embankment are the workmen's cottages (extant) and a reservoir for water supply to the boilers.

The remainder of Bazalgette's account focuses on the engine-house. Its plan – in the form of a Greek cross – was tailored to accommodate the eight enormous beam-engines, each of 142 horsepower with 37-foot beams and cylinders of four-feet six-inches in diameter. These engines, replaced with electric counterparts in 1933, were originally 'arranged in pairs, each arm of the building containing one pair placed parallel to each lengthwise of the arms ... [with] the cylinders ... arranged symmetrically round the centre of the building under the dome'.[37] Bazalgette proceeds to describe the spaces of the building largely in terms of their engineering function, a description that corresponds closely to several of the 52 drawings produced in Bazalgette's office

5.5 An early example of water pumping station design in the engine-house at Kew Bridge, 1838.

5.6 A water pumping station in the guise of a medieval castle: Green Lanes, Stoke Newington, built from 1854 to 1856, and designed by William Chadwell Mylne.

for the building's contractor, William Webster, in 1865.[38] Drawing number 8 (**5.8**), a longitudinal section through the building, provides a case in point; according to Bazalgette's description, the four storeys of the engine-house – two below and two above ground – are arranged so as to accommodate the engines and associated machinery, shown on the left-hand side of the drawing in the two above-ground storeys. The incoming low-level sewer conveys its contents to underground wells, shown in the lowest storey, where they are then lifted by rectangular cast-iron pipes into the sewage pumps. From the pumps the sewage is forced through cast-iron cylinders six-feet in diameter, running along the centre of three of the arms of the building (one of which is shown in the left side of the drawing, along the line marked 'Ordnance') into an air vessel in the centre of the building. From there the sewage is lifted into a cast-iron cylinder (shown on the right of the drawing, just above the line marked 'Datum'), which is carried from the engine-house through the yard into the outfall sewer.

Whilst Bazalgette's description of Abbey Mills is precise in its detail, it tells us very little about either the appearance of, or the reasons for, the lavish exterior decoration or the ornamental ironwork of the interior (seen in the centre of the engine-house and in the upper storey in **5.8**), which obviously added enormous cost to the project. Moreover, Bazalgette's account does not tell us anything about where, in terms of the design, Bazalgette's responsibility ended and Driver's began; indeed, there is no mention of Driver at all.

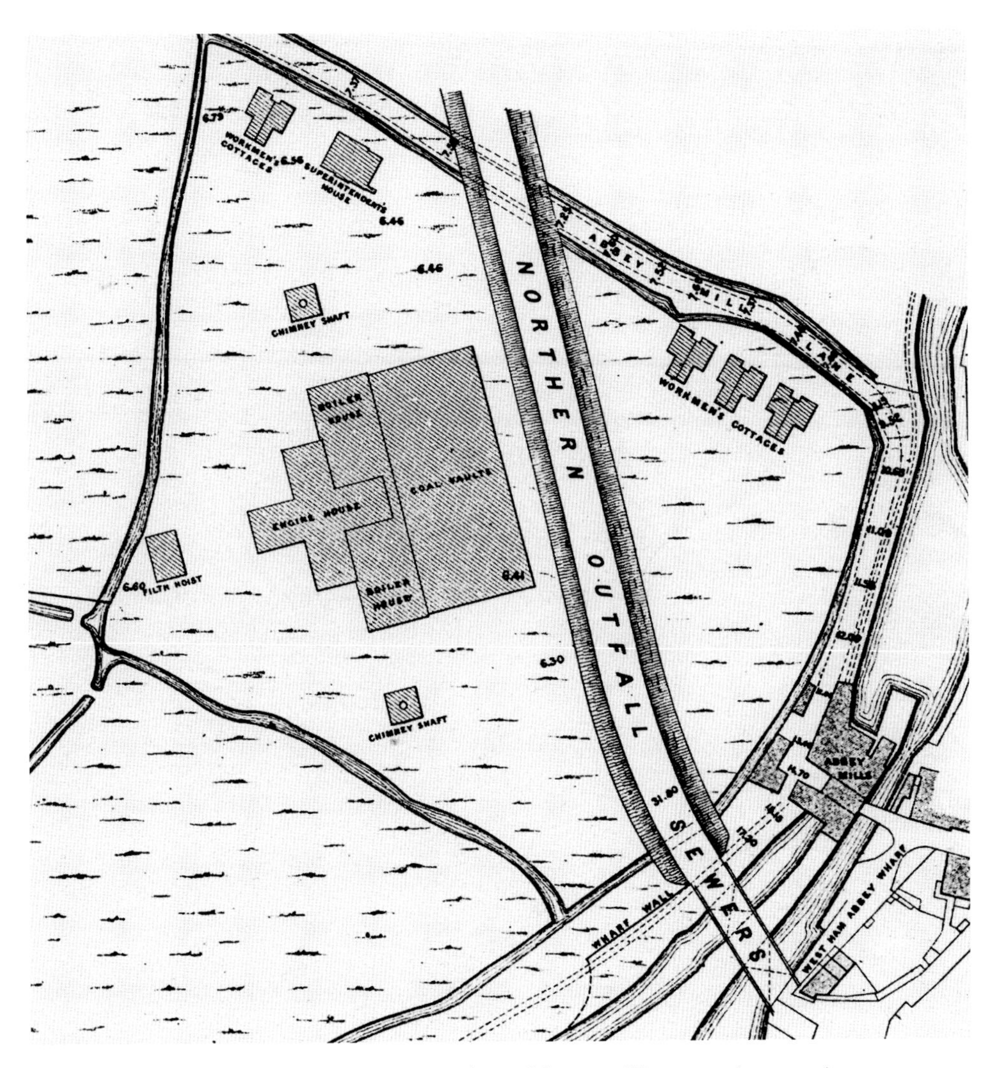

5.7 Extract from drawing no. 1 in the Abbey Mills pumping station contract (1865) showing a plan of the intended works.

The Invisible Architect?

If Bazalgette's account of Abbey Mills sheds no light on Driver's contribution to its design, other relevant documentary sources also yield little information as to his role. The minutes of meetings of the Metropolitan Board of Works show that some payments were made from Bazalgette to Driver: on 10 May 1867 he was paid £42 for 'professional services, *re.* Abbey Mills Pumping Station' as well as £97 for unspecified work during the last two weeks in April.[39] He also received a payment of £8 in July 1864 for 'travelling expenses' to an unknown location,[40] as well as £35 the following year for two watercolour drawings he made depicting the Crossness pumping station.[41] These two drawings were exhibited at the Royal Academy in 1865, under Bazalgette's authorship, despite the fact that they were clearly signed by Driver.[42] Not surprisingly, the architectural press questioned the identity of the architect, but Driver's contribution was not brought to light at the time.[43]

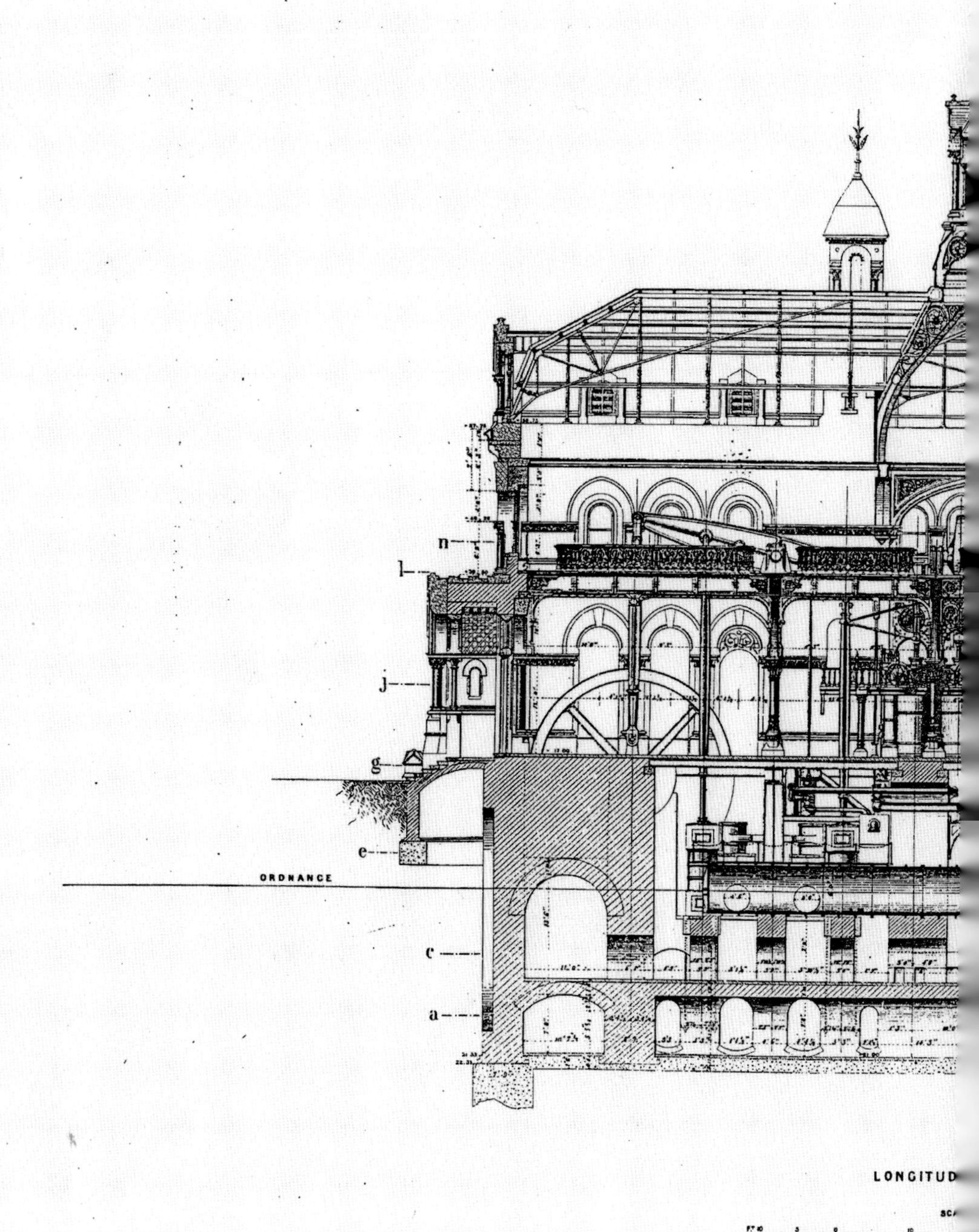

5.8 Drawing no. 8 in the Abbey Mills pumping station contract (1865): a longitudinal section through the engine-house showing the positions of both the steam engines and ornamental ironwork.

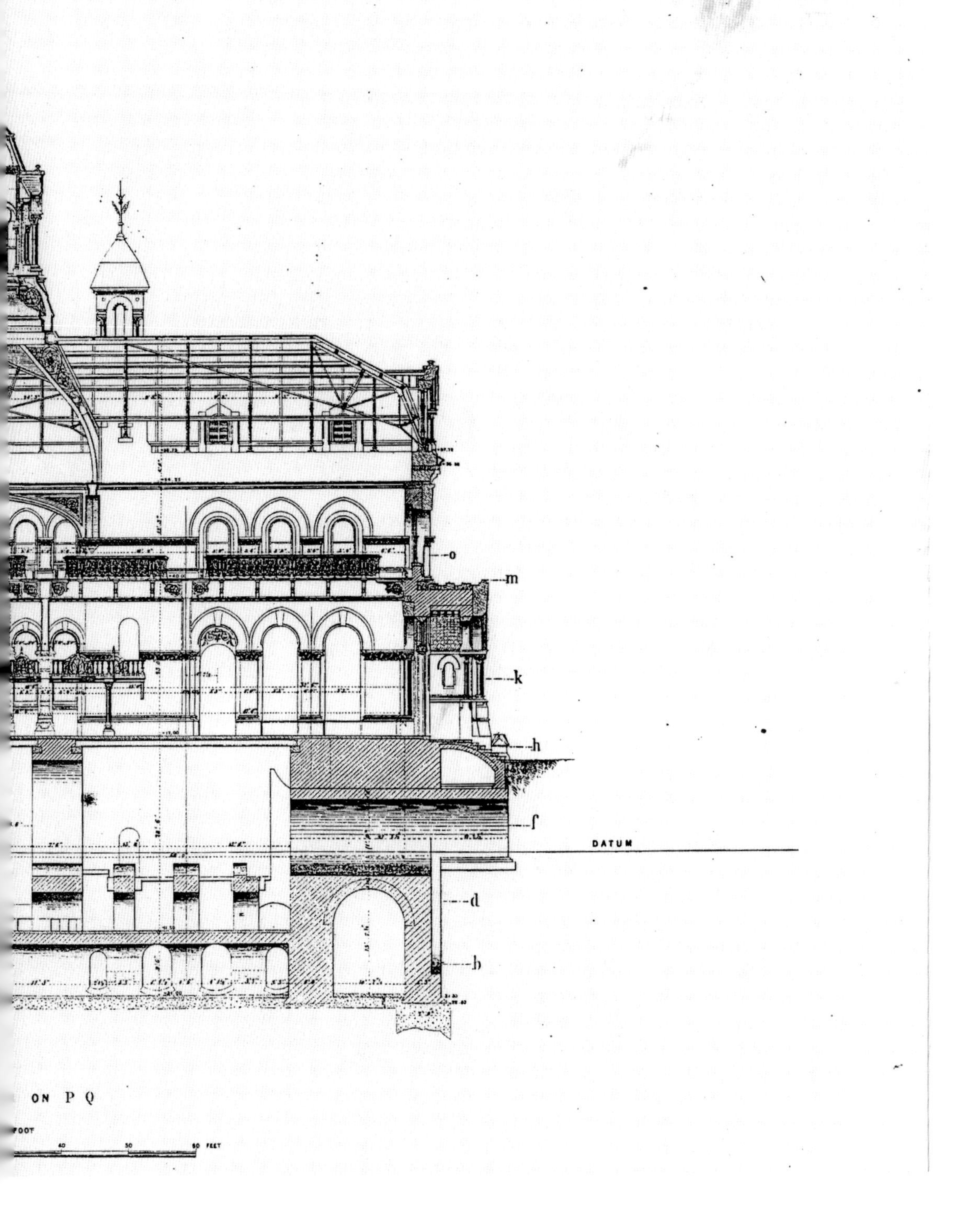

MPING STATION
HOUSE
CONTRACT DRAWING Nº 8.
BUILDINGS &c.
o
m
k
h
f
DATUM
d
b
ON P Q
FOOT
FEET

Other missing evidence regarding the conception of Abbey Mills raises further questions. Unlike any other part of the main drainage system, the Abbey Mills pumping station did not have an official contract associated with it. In August 1865, Bazalgette and his team of draughtsmen had prepared the 52 drawings for the Abbey Mills contract as well as a draft specification,[44] but because the contractor William Webster was already digging the foundations on the site, Bazalgette argued that to submit the contract to the normal tendering process would result in long delays.[45] The Metropolitan Board of Works agreed to Webster undertaking the work, with the result being the absence of a legal contract, only draft specifications for the Abbey Mills buildings, uncertainty about the location of the original drawings, and, most surprisingly, no record in the Board minutes of any debate on the style or cost of the proposed decoration. This lack of information is rendered more mysterious by the fact that, during the similar process of planning the Deptford pumping station in 1859, the Board of Works minutes state conclusively that the civil engineer and architect Thomas Hawksley was requested by Bazalgette to prepare the specification and contract for the building.[46] The absence of any equivalent references to Driver during the preparation of the drawings of Crossness or Abbey Mills leaves many questions unanswered regarding his role in the design process. Without documentary evidence, these questions can only be addressed by examining the building itself and by reconstructing, piece by piece through a close analysis of its architectural details, a more precise picture of Driver's contribution as well as a sense of his architectural vision as a whole.

Exterior: Architecture and Morality

Driver's formative years as an architect-in-training were influenced by important changes within architectural theory and practice in the mid-Victorian period. From the mid-1830s onwards, led by the passionate polemic of A.W.N. Pugin (1812-52) and the ensuing Gothic Revival, architecture became increasingly moralised. Pugin responded to the eclecticism of his day by calling for the purification of architecture and the adoption of a single style based on medieval English Gothic models.[47] Consequently, architectural theory became centred on the question of 'truth', with buildings seen as embodiments of truth and communicators of its meaning. In the late-1840s, John Ruskin (1819-1900) both intensified this perceived moral function of architecture and also broke down the stylistic purity propagated by Pugin, instead reflecting and encouraging the trend to draw inspiration from wider European medieval architecture, in particular examples from Italy. Ruskin's *Seven Lamps of Architecture* (1849) presents clear statements as to the 'virtues' by which architecture might be imbued with a moral sense: that is, 'Sacrifice', 'Truth,' 'Power', 'Beauty', 'Life', 'Memory', and 'Obedience'.[48] Ruskin's advice to architects, expanded in the three volumes of *The Stones of Venice* (1851-53), centred on making such virtues visible in architecture and comprehensible to the observer – the test of which would be a heightened emotional response.[49] Ruskin's emphasis on massiveness, colour and naturalistic ornament reflected his demand that architecture must have both moral and emotional content and, by the mid-1850s, he was the dominating influence on young would-be-architects like Driver.

Despite the fact that Ruskin argued that his architectural virtues (which he associated specifically with an expanded notion of the Gothic) were applicable to all building types, he had little to say concerning the new building types that were a direct consequence of industrialisation, other than to condemn any attempts at making these buildings aesthetically pleasing.[50] However, long before Ruskin's influential polemic, architects and engineers sought to imbue utilitarian buildings with just such a sense of architectural virtue. In the late-18th and early-19th centuries, the predominant use of classical motifs in the design of mills, railway structures and factories were attempts to 'civilize' such buildings and elevate their meaning above mere utility.[51] Architects who took up Ruskin's ideas in the 1850s were faced with a bewildering array of new building types that were a direct consequence of expanding industrialisation and technological development. Despite Ruskin's protestations, many architects, including Driver, began to apply Ruskinian principles to such buildings in order to elevate their status and to develop, according to George Gilbert Scott (1811-78), 'a style which will be pre-eminently that of our own age'.[52] Engineering structures, such as water and sewage pumping stations, were particularly associated with the perceived progressive quality of the Victorian age; the imbuing of sewage pumping stations such as Abbey Mills with a sense of elevated virtue or nobility is not surprissing given the particularly moral character placed on improved sanitation.[53]

Façades

The external appearance of Abbey Mills (**Pl. VI: a**) articulates a strong sense of architectural literacy modelled on Ruskin's principles. The rich use of coloured stones and bricks, both in horizontal bands in the façades and in the window and door arches, is a direct consequence of Ruskin's emphasis on the importance of colour in architecture, which he saw as a major factor in the visual impact of buildings, achieved here through the use of different-coloured building materials.[54] The repeal of brick tax in 1850 and advancements in both the mechanisation of production and in the techniques of colouring building materials provided a practical impetus for the implementation of such constructional polychromy. Likewise, the façades – with arcaded fenestration on two storeys separated by a bold string course, the intricate and deeply-recessed cornice, and the double-pitched Mansard roof – follow Ruskin's dictate that buildings should appear massive and that this should be evident to the eye of the observer by means of a 'continuous, visible bounding line'.[55] The use of arcaded fenestration was a popular device employed by architects in the 1850s and 1860s mainly in the design of public, commercial and industrial buildings.[56] Drawn from north Italian medieval models (particularly Venice), celebrated by Ruskin and often illustrated in the pages of the *Builder*,[57] architects justified such arcading on both associational and practical grounds: in terms of association, the status ofVenice as a mercantile city was seen as having an equivalent in the industrial cities of England, particularly London, Bristol,

5.9(a) Windows in the first floor of the west bay of the Abbey Mills pumping station, showing Driver's use of both round-headed and pointed arches.

and Manchester; in terms of function, multiple rows of windows provided both additional light and the possibility of multi-storey façades. Arcaded fenestration, like the increasing use of coloured building materials, was also made more economically viable by the repeal of glass tax in 1845 and window tax in 1851.

Abbey Mills represents a rather late application of Ruskin's ideas: by the end of the 1860s some architects were questioning the excessive use of colour in architecture that had resulted from very liberal interpretations of Ruskin's ideas by others in their profession.[58] However, viewed in the light of Driver's previous projects, Abbey Mills can be seen as the climactic point in his early career – a career that only began in the late-1850s. The façades at Crossness (**5.3**) show a similar ordering of windows as well as a steeply-pitched roof punctuated by louvers, while Driver's design for the railway station and viaduct at London Bridge (*c.*1866: **Pl. VI: b**), with its imposing triplets of round-headed blind arches, effectively conveys a sense of massiveness by repetition.[59] Driver's station at Battersea Park (1866: **Pl. VI: c**), with its three-storey

5.9(b) Windows in the dormer level of the end bays of Abbey Mills from drawing no. 27 in the Abbey Mills pumping station contract (1865).

façade with polychrome, round-headed and deep-set arches on the ground and second floors, corresponds closely to the effect conveyed at Abbey Mills. Another striking feature of the façades at Abbey Mills is the skilful integration

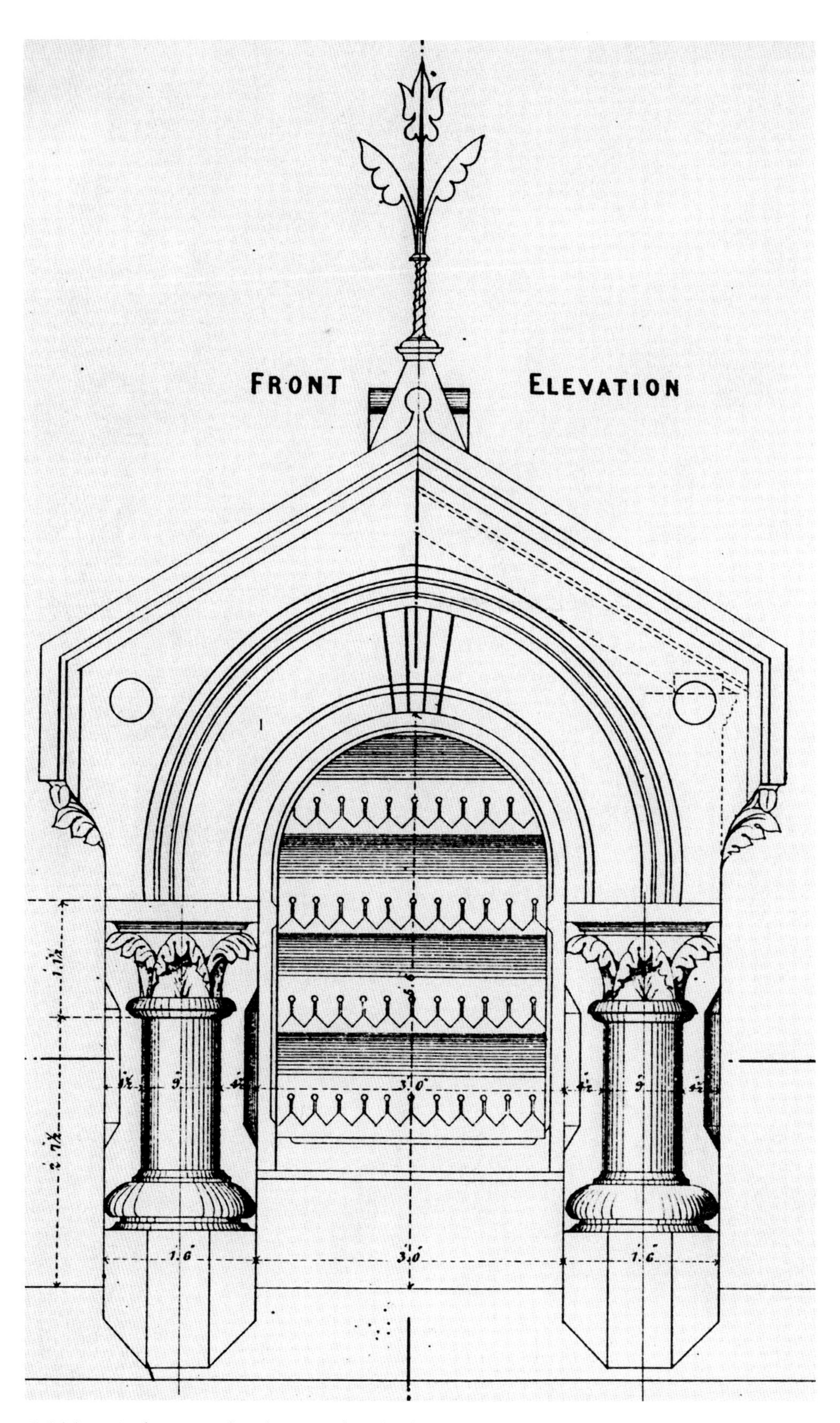

5.9(c) Windows in the dormer level of the side bays of Abbey Mills from drawing no. 27 in the Abbey Mills pumping station contract (1865).

5.9(d) Window from an old house in Verona, Italy, as drawn by G.E. Street in his *Brick and Marble in the Middle Ages* (1855), p. 94.

of the spiral cast-iron drainpipes into the stonework – a feature resembling spiral colonettes found at the corners of medieval and Renaissance Venetian palaces, which were illustrated in the *Builder* in 1851.[60] Driver employed this motif previously in his stations at Battersea Park and Peckham Rye and probably derived the idea from its use in the façades of the iron foundries of Walter Macfarlane in London and Glasgow, whom Driver used to manufacture most of his ironwork in the 1860s.[61] Driver drew upon these medieval Italian sources for both aesthetic and practical reasons: if the cast-iron drainpipes were given architectural 'virtue' through these more elevated sources, this virtue was achieved without compromising their prosaic function. Such a marrying of function with a particular kind of medieval historicism follows the example of high-profile architects such as George Gilbert Scott, George Edmund Street (1824-81) and Benjamin Woodward in the 1850s and 1860s, who responded

5.9(e) Window in the first floor of Leatherhead railway station, built from 1865 to 1867 and designed by Charles Driver.

to Ruskin's call to 'universalise' the application of an eclectic Gothic style to secular buildings. The motifs in the façades at Abbey Mills represent an attempt to elevate a building with a prosaic function by drawing on Ruskinian ideas of 'nobility' – ideas that were familiar to Driver, as evidenced in his concurrent architectural projects.

Windows and Doors

If the exterior appearance of Abbey Mills suggests a north Italian or Venetian derivation, a more detailed examination of the forms of the windows and their arches indicates a much broader field of influence that includes ideas unique to Driver's work. Three main window types are employed in the Abbey Mills engine-house: the first (**5.9: a**) is a window and arch-type commonly used by architects in the 1850s and 1860s – that is, windows framed by arches with rounded intrados and pointed extrados. According to Ruskin, this represented the strongest and most 'noble' form of arch because it visibly displayed a building's sense of 'masculinity'.[62] George Edmund Street and George Gilbert Scott also saw this type of arch as achieving a synthesis between Classical and Gothic forms,[63] producing what Street termed 'hybrid' arches that he observed in north Italian medieval buildings.[64] The use of alternating red, black and sandy-coloured bricks in these arches correspond closely to medieval windows in Verona illustrated in Street's 1855 book describing an architectural tour of north Italy (**5.9: d**). The prolific use of this type of arch in the façades at Abbey Mills suggests that synthesis – as well as strength – is a key theme being explored by the architect.

This is confirmed by the inclusion of two other window forms in the exterior of Abbey Mills. The larger windows – seen in the ground floor of the workshop, in the central bay above the main doors, and most prolifically in the lantern above these – combine a host of different elements, including medieval arches, a Renaissance form of window in the lower part of the building (**5.9: b**) and a highly individualised arch form in the dormer and lantern windows (**5.9: c**). The cast-iron frames of the lantern windows, in the form of two small round-arches supporting a roundel, are derived ultimately from Italian models, but executed here in a modern material.[65] This window type, used in many other industrial buildings contemporaneous with Abbey Mills, allows extra light and air into the building, but also serves to synthesise Renaissance and medieval forms.[66] The overtly massive four-sided stone hoods surmounting these windows in the dormer level and in the lantern, give further emphasis to the notion of strength seen in the other windows. In addition, this highly individual form, unique to Driver's work and also used at Leatherhead station (1865-67: **5.9: e**), further reinforces the sense of synthesis evidenced in the varied forms of windows and arches seen throughout the exterior. Such synthesising represents an attempt to combine both historical and individual forms in a personalised hybrid style.

The four main doors and porches at Abbey Mills (**Pl. VII: a**), arranged symmetrically in the centres of the end bays of the engine-house, display a remarkable confluence with the window forms. The form of the porches in

5.10(a) The original design for the lantern at Abbey Mills from drawing no. 15 in the 1865 contract.

5.10(b) The completed lantern as seen in the *Illustrated London News*, 15 August 1868, p. 161, showing the raised section of roof added during construction.

5.10(c) Deane and Woodward's original design for the laboratory at the University Museum, Oxford, from the *Builder*, 7 July 1855, p. 319.

particular also points to the high level of Driver's involvement in the design of Abbey Mills: with their quadripartite hoods framing richly-coloured archivolts with rounded intrados and pointed extrados (both united by an extended keystone), their double columns on double plinths, and the curious triangular adjuncts to the pedestals, these porches are unique to Driver's oeuvre, and he used them in a very similar form at Leatherhead station (**Pl. VII: b**) in 1865. Driver's application of the quadripartite form in the windows and doors at Abbey Mills and Leatherhead is unique, but its ultimate derivation may come from Italian models such as the 13th-century monument of Antenor in Padua mediated through his more famous contemporaries: Street used it in his tower of the church of St James the Less (1859-61) in Pimlico.

Lantern

The strong sense of synthesis in the exterior of Abbey Mills reaches a visual climax in the octagonal dome or lantern that spans the crossing at the centre of the engine-house (**5.10: a**). Constructed from a mixture of pre-fabricated cast and wrought iron and timber, and painted to harmonise with the stone and brick exterior, the lantern presents a formidable array of possible precedents. Architectural historians have variously described it as a 'dome like that of an Eastern Orthodox church',[67] a 'Byzantinesque lantern',[68] 'a Slavic dome',[69] or 'almost Russian in appearance'.[70] However, in relation to Abbey Mills, the most recent and influential attempt to combine historical forms with modern materials was the University Museum in Oxford (1855-60) by Benjamin Woodward and Thomas Deane but permeated with a great deal of influence from Ruskin.[71] Significantly, one of the most original and striking features of the University Museum is the octagonal laboratory (**5.10: c**), set apart from the main building, and directly modelled on the 14th-century Abbott's Kitchen at Glastonbury Abbey (**5.10: d**). In effect, the architects transformed a medieval building into one that could accommodate the requirements of modern science and technology. The main drainage pumping stations, in terms of their design, would have required a similar direct relationship between historical models and modern technology (in this case the large-scale pumping machinery). In this regard, the Abbott's Kitchen at Glastonbury, in the light of its successful adaptation in the Oxford Museum laboratory, would have seemed an appropriate model for their design.

There is another important factor that would have reinforced the appropriateness of the Abbott's Kitchen as a model for Abbey Mills: the pumping station was built on part of the site of another ancient abbey in West Ham, the abbey of Stratford-Langthorne, largely destroyed in the reign of Henry VIII.[72] Indeed, in 1897, Edward Walford located this ancient abbey in direct relation to the Abbey Mills pumping station, commenting that the latter occupied 'about seven acres of the ground once covered by Stratford Abbey'.[73] This lends weight to the possibility that Driver, given his high degree of visual literacy, was aware of proximity of the ruined abbey of Stratford-Langthorne to the pumping station and consequently was able to justify the references to Glastonbury Abbey in the design of the lantern on associational grounds.

5.10(d) The model for Deane and Woodward's laboratory in Oxford: the Abbott's Kitchen at Glastonbury Abbey, built in the 14th century.

The original double lantern of the Abbott's Kitchen (**5.10: d**), with its three-level pitched roof, compares closely, in terms of its geometry, proportions and function, with the lantern at Abbey Mills (**5.10: b**). The four turrets that flank the Abbey Mills lantern also resemble the compositional arrangement of the reconstructed 'chimneys' in the laboratory of the University Museum, Oxford. Furthermore, the Abbey Mills lantern was altered during construction, increasing its resemblance to the Abbott's Kitchen. The changes made are clearly illustrated by comparing **5.10(b)** (the finished product as shown in 1868) with **5.10(a)** (the original design depicted in the 1865 contract drawings). For some unknown reason, the height of the lantern was increased during construction; it was raised up on vertical iron plates with a section of pitched roof added above the level specified in the contract. The vertical plates were then covered in slates and ornamented with a zigzag stencilled pattern. The addition of ornamental cast-iron gargoyles (for drainage) also adds a medieval character to the lantern – one not proposed in the original contract.

However, the most convincing evidence for the validity of this comparison is the perceived fusion of function and form in these kitchen structures, explicitly drawn upon in contemporaneous commentary on the University Museum laboratory. Despite the fact that its ventilating function is never explicitly stated by Bazalgette, the lantern at Abbey Mills – with its symbolic associations with medieval kitchens, its position in the centre of the building surrounded by the sixteen louvered dormer windows in the roof (explicitly for ventilation purposes),[74] and the alterations made during construction – suggests that its function was to both draw up the fumes generated by the engines inside the building and also to bring in a copious amount of light. These twin tenets of ventilation and light were key principles not only in the design of the University Museum laboratory in Oxford but also in the wider sphere of sanitary improvement in the mid-Victorian period;[75] the functional and symbolic role of the lantern, made clearer in the alterations carried out during its construction, closely articulates these principles.

Chimneys

The importance of ventilation in the function – symbolic or otherwise – of the Abbey Mills lantern is also evident in the twin chimneys that, until the Second World War, stood either side of the engine-house – their function being to remove the smoke generated by the boilers (**5.1**).[76] Rising to a height of 209 feet, the chimneys comprised tapering octagonal shafts placed on monumental square bases – the entrances of which corresponded in style to the doors and windows of the engine-house. The chimney shafts were decorated with a series of interlocking hexagons of coloured brick and were capped by an 'ornamental cast-iron roof, pierced for the egress of the smoke'.[77] The earlier, equally striking, 178-foot chimney at Crossness seen in the background of **5.3** – with its polychromatic bands of red and white brick and stone, corbelled cap and cast-iron canopy crowned with beaten wrought ironwork – was modelled on medieval Italian campaniles, particularly the Torre del Mangia (1325-44) in Siena. Such 'dressing up' of chimneys to resemble campaniles was another consequence of Ruskin's celebration of medieval Italian architecture, in particular the capacity of their towers to convey a strong sense of power and strength.[78] However, Ruskin was appalled by the increasing use of the campanile form for industrial chimneys: unlike the architects who readily applied his theories to the design and decoration of chimneys, Ruskin himself viewed the modern chimney form as an essentially degraded product – and most obvious sign – of industrialisation, without moral value because incapable of embodying his notions of architectural virtue.

In the 18th century, chimneys were short, normally square in plan, and attached to small industrial buildings such as mills and factories. By the mid-19th century, as these existing building types became larger and new building types emerged, chimneys became both taller and more numerous. The chimney attached to the Palm House at the Royal Botanic Gardens, Kew, built by Decimus

5.11(a) The chimney serving the Palm House at the Royal Botanic Gardens, Kew, built in 1847 by Decimus Burton and one of the first to be modelled on an Italian campanile.

5.11(b) Robert Rawlinson's fantastical array of industrial chimneys as seen in the *Builder*, 25 April 1857, p. 23.

THE PROPOSED ELLESMERE MEMORIAL, WORSLEY, LANCASHIRE.—MESSRS. DRIVER AND WEBBER, ARCHITECTS

5.11(c) Charles Driver's model for the Abbey Mills chimneys, the Ellesmere Memorial in Worsley, Lancashire, as seen in the *Builder*, 6 November, 1858, p. 747.

Burton in 1847, was probably the first time that a chimney was designed to resemble an Italian Romanesque campanile (**5.11: a**).[79] Situated 500 feet from the iron and glass Palm House, it stands in stark juxtaposition to its parent building with no attempt made to aesthetically integrate the two. However, by the early-1860s, concerted attempts were made to incorporate chimneys into a unified architectural programme. Robert Rawlinson (1810-98), an architect and engineer who specialised in pumping station design, dealt with the subject in detail in his *Designs for Factory, Furnace and Other Tall Chimney Shafts* (1858), with extracts reproduced in the *Builder* in 1857 (**5.11: b**).[80] Rawlinson's treatise praised Italian campaniles and eastern towers and minarets and he regarded them as appropriate models for the enhanced architectural treatment of modern chimneys, a view that would set the precedent and justification for a generation of fantastical chimneys that would follow in the 1860s.

If the campanile-like chimney at Crossness (**5.3**) conforms closely to Rawlinson's examples, the twin chimneys at Abbey Mills (**5.1**) present a more original design that synthesises a host of different elements. Forming a striking aesthetic unity with the engine-house, the appearance of the chimneys prompted some observers to describe the ensemble as resembling a mosque or Chinese temple.[81] Certain elements follow Rawlinson's examples, such as the massive bases of the chimneys and the spurs marking the transfer of the square base to the octagonal shaft. However, the tapering form of the shaft, Gothic canopy and hexagonal decoration are Driver's own invention, closely modeled on one of his first architectural projects: the Ellesmere memorial in Worsley, Lancashire, designed in 1858 in conjunction with John Huish Webber and illustrated in the *Builder* in the same year (**5.11: c**).[82] The caps of the chimneys at Abbey Mills are a simplified adaptation of this design, constructed in cast iron, and the interlocking hexagonal decoration on the shaft reflects the individualistic use of tiles in the shaft of the memorial. This highly personalised treatment of the chimneys – incorporating elements from Rawlinson's designs and from Driver's own oeuvre – mirrors the treatment of the lantern and windows, outlined above; once again it suggests both Driver's close involvement in the design of Abbey Mills and also his comprehensive approach – an approach characterised by the combination of eclectic historicised forms and personal elements into a unified whole.

Decorative Details

A closer analysis of the decorative details employed in the exterior of the engine-house serves to confirm Driver's close involvement in the design of Abbey Mills and highlight the presence of a comprehensive decorative programme. The prolific use of coloured encaustic tiles, both in the cornice and in the recesses under the windows of the engine-house, was, by the late-1860s, a common method by which architects introduced colour into buildings. Produced by the Minton Company in Stoke-on-Trent, founded in 1835, these decorative tiles were used in prolific numbers by many architects in the 1850s and 1860s: Driver used them in the exterior of Dorking station (*c*.1865)[83] and in again in the interior of the chancel of St Mary's Church (1868) in

5.12(a) Encaustic wall tiles in the chancel of St Mary's Church, Warkworth (1868) showing Driver's liberal use of a stylised flower motif.

Warkworth, Northamptonshire, one of his first projects as an independent architect after Abbey Mills.[84] The motifs employed in these tiles correspond closely to those used at Abbey Mills. **Pl. VIII(a)** shows tiles in the recesses below the ground-floor windows at Abbey Mills and repeated all around the building; **Pl. VIII(b)** illustrates similar, if more elaborate, tiles employed in the floor of the new chancel at St Mary's in 1868. Both designs may have been drawn from medieval floor tiles depicted in Owen Jones's influential design sourcebook *The Grammar of Ornament* (1856: **Pl. VIII: c & d**).

5.12(a) shows another tile design employed by Driver in the walls of the chancel at St Mary's – a further indication of the level of correspondence between Abbey Mills and Driver's other works. The stylised five-petal flower

5.12(b) Driver's flower motif and symbol of the London, Brighton and South Coast Railway, in a keystone in a window arch at Battersea Park railway station (1866).

motif compares closely with a decorative feature employed in many of Driver's other buildings: that is, keystones decorated with stencilled flower motifs, used in all of his stations for the London, Brighton and South Coast Railway (**5.12: b**). This motif is repeated at Abbey Mills, in both the enlarged keystones of the arches in the second-floor windows (**5.12: c**) and also in the elaborate brass hinges of the four main doors (**5.12: d**). It was a motif that Driver was to use again in 1869 in the archivolt of the main entrance to the Horton Infirmary in Banbury (**5.12: e**). Such a consistent use of this motif in Driver's oeuvre indicates both his involvement in the decorative details of Abbey Mills and also the comprehensive quality of his decorative scheme.

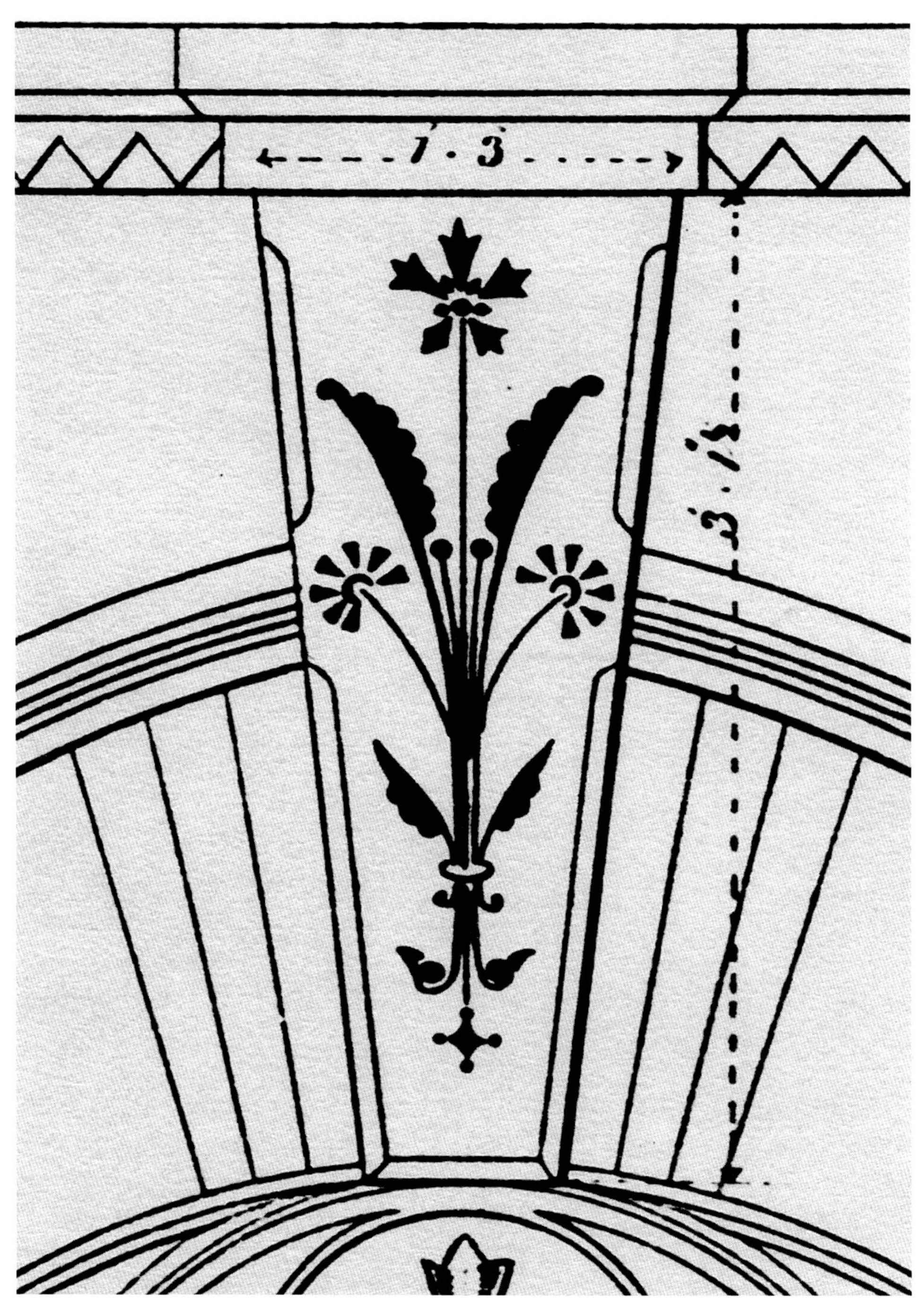

5.12(c) Decorated keystone in the second-floor window arch at Abbey Mills as depicted in drawing no. 24 of the contract for the building (1865).

Interior: Engineering and Art

When the *Illustrated London News* depicted Abbey Mills on the occasion of its opening in 1868 (**5.1**) it showed two dramatically different views of the building: one showing the flamboyant exterior, the other the interior showing the extraordinary decorative ironwork and parts of the enormous steam engines housed therein (seen in the left and right foreground). One reason for such a

5.12(d) One of the many ornamental brass hinges in the panels of the four pairs of wooden doors at Abbey Mills.

dramatic contrast between exterior and interior lies in building regulations that were introduced in London in 1855, which actively discouraged exposed – that is external – cast-iron construction, on the basis of its perceived risk as a fire hazard and the danger of oxidisation and decay. Another reason, however, relates to the character of pumping stations themselves, a character shared with other new building types in the Victorian period. Railway stations, markets, factories and warehouses all required large, undivided internal spaces that were only achievable through a structural use of cast iron, which was much stronger in compression than any traditional building material; by contrast, because cast iron provided the main internal structural support, the exterior of these buildings allowed for a more conventional stylistic treatment in traditional building materials. The engine-houses of pumping stations are a case in point; if the gigantic steam engines housed in the interior spaces needed large undivided spaces both to accommodate their bulk and to allow easy access to their constituent parts, then the exterior – a decorative shell or more properly a 'house' – could be treated very differently without compromising the building's engineering function. Nevertheless, the interior of Abbey Mills is clearly much more than simply a 'functional' space made possible by the structural use of iron; it is also extravagantly and, at first glance, unnecessarily, embellished with elaborate decoration. This illustration does not reveal the reasons for such decorative flamboyance and how it relates to the building's functional requirements; the rest of this chapter will offer possible answers to these questions.

5.12(e) Decorative stones in the archivolt of the main entrance to the Horton Infirmary in Banbury, Oxfordshire, built from 1869 to 1872 and designed by Charles Driver.

Ruskinian Iron

The question of the validity, or otherwise, of iron in creating a new architectural style was perhaps the single most important issue confronting architects in the mid-Victorian period, especially after the construction of the Crystal Palace in 1851. This building not only thrust the question of iron architecture into the public sphere but also made a decisive impact on the way in which young architects like Driver would go on to attempt to solve the aesthetic problems posed by this new material. Many Victorian architects viewed the aesthetic quality of the Crystal Palace as a direct attack on the central tenets of architecture itself. For John Ruskin, their most eloquent spokesman, these offences centred on the question of 'truth' in architecture and on the problematic visual appearance of iron when used as a 'naked' structural or decorative material. According to Ruskin, iron offended the eye because of qualities inherent in its materiality; for him, the thin iron columns that supported the Crystal Palace did not appear massive in the manner of traditional materials such as brick or stone. When discussing the use of iron for decorative purposes, Ruskin was even more strident: he condemns cast-iron or machine-made ornament as a 'lie ... an imposition, a vulgarity, an impertinence, and a sin'.[85] Such strongly moralistic language stemmed from Ruskin's hatred of industrial society itself, with its division of labour and the resultant absence of handiwork in a mechanised production process. If Ruskin compared cast iron to the corrupting influence of an immoral person, it is because he objected to the very basis of the society that produced it.[86]

Such a negative attitude towards iron might seem unlikely in an architect like Driver, who from his very first projects used it for new building. Yet, in his architectural thinking, as voiced in papers presented to the R.I.B.A. and the Society of Mechanical and Civil Engineers in the 1870s, Driver engages with Ruskin's views and adopts many of his principles without compromising his embracing of iron as a valid constructive and decorative material. Two papers presented to the Society of Mechanical and Civil Engineers in 1874 and 1879 make liberal references to Ruskin's writings. In his 1874 paper, 'Engineering, its Effects upon Art', Driver uses Ruskin's own words when he states that 'the real source of all inspiration in art (or design) is to be found in nature' and that art is distinguishable from mere 'manufacture' by the 'manifesting of *human design* and authority' (emphasis in original).[87]

By quoting Ruskin, Driver seeks to invest his own lecture with the weight of Ruskin's considerable authority. At the same time, however, he directly challenges Ruskin's negative views on mechanical reproduction and consequently the use of cast iron in architecture. Discussing 'the good that engineering has done to art', Driver argues that the reproduction of art by machines does not change the status of a work of art; rather it distributes its beauty to the many rather than the few. He cites the example of a gardener who disseminates a plant by making cuttings from a single source: for Driver 'the plant or flower has not changed, it is just as beautiful as ever, but instead of pleasing the few it pleases the many'.[88] Such an argument, made against the perceived elitism of critics like Ruskin, prepares Driver's audience for his defence of cast iron, taken up more fully in a paper presented to the R.I.B.A. in 1875.[89] In stark contrast to Ruskin, Driver views cast iron as good in itself, a 'valuable friend, equal indeed to most other building materials and superior to some; invaluable both for constructive and decorative purposes'. According to Driver, the reason behind Ruskin's and other architects' negative view of iron lay in their lack of engagement in engineers' projects, and 'not because architects are jealous of the success of the engineers, but rather because of the disgust they feel at the inartistic result of their labours'.[90] Using the analogy of the human body, Driver argues that iron should not be hidden from view by being encased in plaster, as Ruskin had suggested, but should be fully exposed to the eye:

> ... for the flesh is not put on to hide skeleton merely, as the lath and plaster is put on to hide the iron. The skin and flesh and muscle perform most important and necessary functions ... but this is not the case as regards the lath and plaster casing to iron; strip it off and the building stands as well as ever.[91]

For Driver, acceptance of iron in its 'naked' form requires the retraining

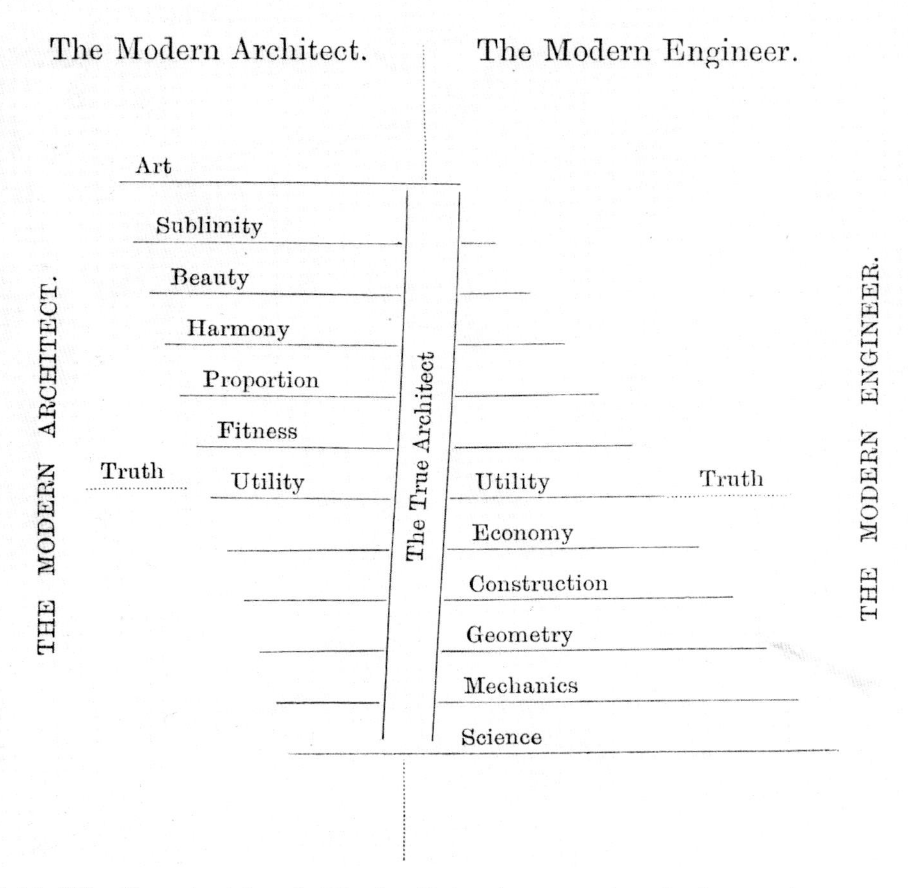

5.13 'The True Architect', Charles Driver's eccentric solution to the Victorian separation between architecture and engineering, published in Driver's paper 'Engineering, its Effects upon Art', *Transactions of the Civil and Mechanical Engineers' Society* (1874), p. 9.

of the observer: the 'eye may require some amount of Education before it becomes accustomed to the use of iron and its employment in connection with other materials'. He asserts his hope that architects, with such education and a willingness to fully engage with iron, 'might produce designs for iron which though not perhaps in accordance with any existing particular style, shall yet harmonize, even perhaps by contrast, with them'.[92] By using examples drawn from nature, Driver attempts to argue the case for iron in almost Ruskinian terms; in his consistent use of iron alongside traditional building materials in his architectural work – in particular in the interior of Abbey Mills – Driver invests iron with the very principles that Ruskin put forward to justify his objection to it.

It is not known what Driver's audience of engineers in 1874 and 1879 would have made of his lengthy references to Ruskin's writings, as the discussion that followed these lectures are not recorded. Nevertheless, that which accompanied his lecture to the R.I.B.A. in 1875 was published and presents an instructive example of how Driver's views on iron compare with wider debate on this issue. On that occasion his audience was made up of engineers, iron manufacturers, architects and architectural historians: a cross-section of the main parties involved in the debate on iron that had been going on since the 1840s in similar discussions and in the pages of the architectural and engineering press. In this case, the architect Thomas Chatfeild Clarke raised the fiercest objections to Driver's defence of cast iron, arguing, in terms that mirrored Ruskin's, that the 'spider-work constructions' like the Crystal Palace had no beauty or sense of 'repose'.[93] The engineer John Dixon countered Clarke's negativity, arguing, like Driver, that architects needed to be educated about the material qualities of iron before anything 'worthy' could be produced.[94] It was, however, the iron manufacturers Walter Macfarlane, Ewing Matheson and Francis Skidmore who responded most enthusiastically to Driver's passionate defence of cast iron; they all argued that architects needed educating about the manufacturing process in order to achieve the best results.[95] Also present in the audience were two architects who were more ambivalent in their response. George Aitchison, architect of Leighton House (1877-78), expressed pessimism about the possibility of cast iron ornament ever being beautiful, not because it was immoral, but because iron was expensive, liable to rust, and functioned most successfully in a structural, rather than decorative, capacity.[96] Not surprisingly, Driver brought the discussion to a close rather abruptly, giving thanks to all who contributed, but no doubt a little exasperated by the lack of consensus of opinion in his favour.

'The True Architect': Unifying Architect and Engineer

Aitchison's pessimistic response to Driver's embracing of cast iron is surprising, given that he had earlier expressed more positive hopes for its use in architecture in a lecture he had delivered in 1864.[97] Yet, by the time Driver presented his own paper in 1875, many of his fellow architects and theorists had given up hope of iron producing a new style of architecture suited to the Victorian age. Rather, Driver's optimism in this regard reflects the attitude of an earlier generation of designers: those associated with Sir Henry Cole (1808-82), known as the Cole Group, who had been those most active in the rebuilding of the Crystal Palace at Sydenham in south London from 1852 to 1854. Among these the architects Owen Jones (1809-74) and Matthew Digby Wyatt (1820-77) were the main spokesmen. For Jones and Wyatt the Crystal Palace was a powerful 'symbol of progress', suggesting the possibility that iron might produce an entirely new style of architecture in the future, reflecting the 'character of the age', that, according to Jones, was defined by a new 'religion of science, commerce and industry'.[98] Jones's attempts, in the 1850s, to use iron in combination with other materials reflected his impulse to embrace this new religion.[99] Wyatt was equally committed, publishing articles praising the qualities of iron and, with Isambard Brunel, experimenting with the creation of a new iron style in the interior of Paddington station (1852-54).[100] At the heart of these experiments, voiced most strongly by the architectural historian James Fergusson, is the hoped-for union of architecture and engineering where the creation of a future new style lay in the hope 'that the engineers may become so influential as to free the Architects to adopt their principles'.[101]

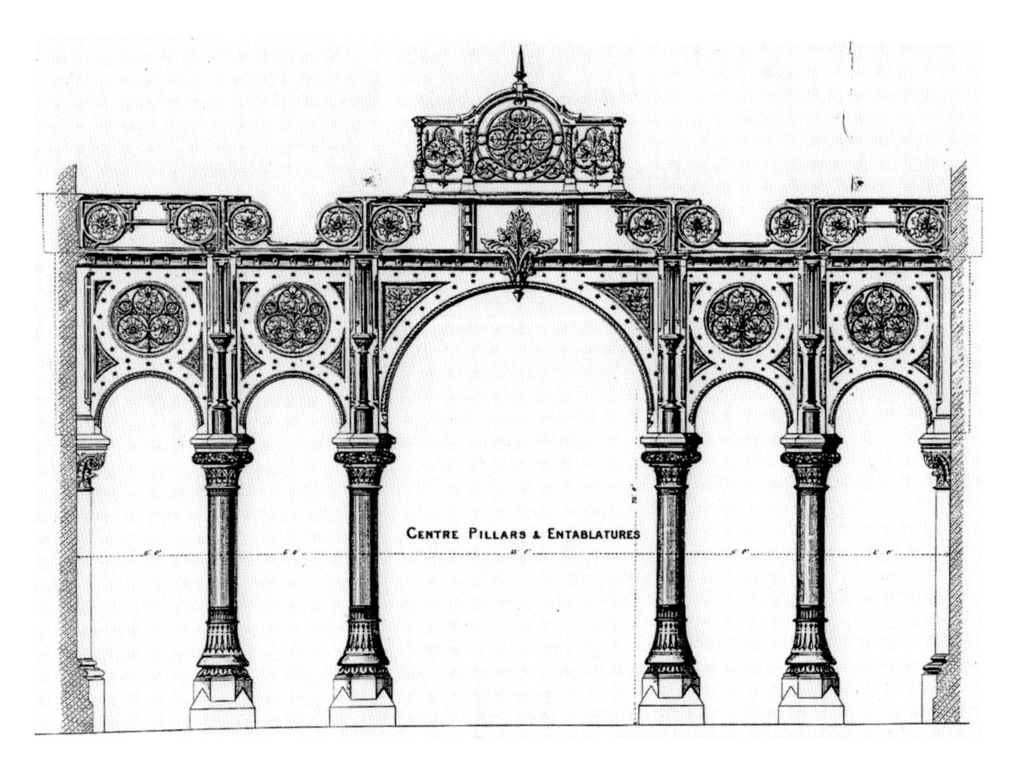

5.14(a) Elevation view of part of the cast-iron octagonal structure in the centre of the interior of Abbey Mills, as seen in drawing no. 12 of the contract for the building's machinery (1864).

5.14(b) Elevation of the Byzantine Court in the Crystal Palace at Sydenham (1856) published in Matthew Digby Wyatt and John Burley Waring's *The Byzantine and Romanesque Court in the Crystal Palace* (1854), p. 52.

Driver, in his published papers, mirrors these hopes, but is more guarded than Fergusson in any celebration of the influence of engineers on the architectural profession. Like most Victorian architects, Driver shared the belief that building and architecture were two distinct areas – architecture being mere building made 'artistic' chiefly through decoration. Whilst praising engineers for their 'honest construction', Driver sees their inability to think 'artistically' as resulting in 'chaos, as regards to Art'. The solution, for Driver, lies in 'Art' being '[t]he master directing and guiding Engineering into the right paths' – that is, the reassertion of the primacy of architects over engineers in order to 'educate' engineering.[102] The achievement of such a reconciliation is seen to rest in a figure that Driver terms 'the true architect', who emerges out of his theoretical fusion of both professions.[103]

In order to illustrate his proposal more fully, Driver included, in his 1879 paper, a diagram (**5.13**) showing the 'qualifications which the engineer and architect should have in common' with 'the true architect' (perhaps a self-portrait of Driver himself) having 'an equal measure of each'. The diagram indicates, by means of horizontal lines, the different qualities pertaining to architects and engineers, with the respective lengths of the lines corresponding to the measure of each quality. With eight qualifications, the modern architect is seen first and foremost as an artist with very little 'science', while the engineer, with his seven qualifications, is primarily a scientist with 'no art' whatsoever.[104] The true architect mediates in all of these categories, and is represented by a strangely off-centre line in the middle of the diagram, with the qualifications of 'truth' and 'utility' providing the central axis around which the other characteristics happily coexist. Such an idealistic vision, expressed in this peculiarly lopsided diagram, represents Driver's highly individual solution to the problem of how to unite architectural and engineering practice.

Driver's insistent emphasis on the fusion of the architectural and engineering professions and his synthesis of conflicting views on iron runs counter to the general tenor of the debate on iron in the 1870s, clearly demonstrated in the discussion that followed his lecture to the R.I.B.A. in 1875. In one exchange, the engineer John Dixon lamented the fact that architects had little knowledge of the material properties of cast iron, which resulted in designs that were not 'worthy ... of the profession',[105] while the architect Thomas Chatfeild Clarke responded by castigating the 'hideous monstrosities' erected by engineers, such as the new London Bridge (1824-31), designed by John Rennie and built by his son. In the middle of Clarke's tirade, Dixon interrupted and angrily stated that 'Engineers do not profess to be architects'.[106] Such arguments demonstrate the level of hostility that existed between both professions, which had intensified after the construction of the Crystal Palace in 1851 and must have seemed intractable by the 1870s. Driver's idealistic vision of the fusion of both professions and his synthesis of their opposing views, as expressed in 1879, is remarkable given such a situation, but more remarkable still was his ability to translate this vision into practice, expressed most clearly in his treatment of the ironwork in the Abbey Mills pumping station.

Octagon

Directly beneath the lantern in the interior of Abbey Mills is a striking octagonal arrangement of twelve ornamental cast-iron columns and entablatures, filling the central area of the building on the ground floor (**5.14: a**). This octagonal structure is developed from a similar feature employed at Crossness, but also follows the precedents set by the interior spaces of certain

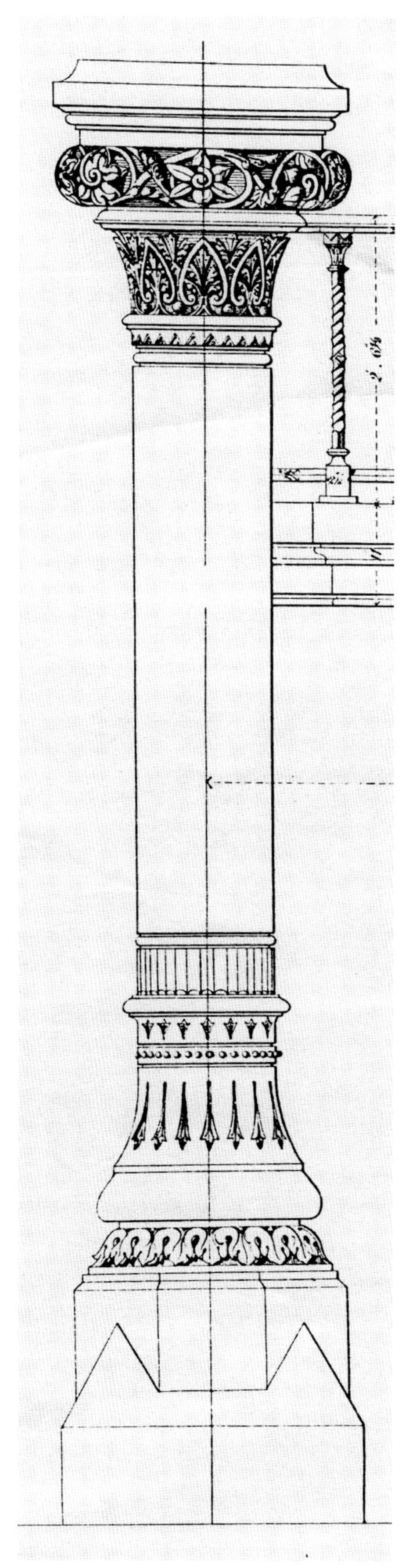

(*left*): 5.15(a) One of the supporting cast-iron columns in the octagon at Abbey Mills as seen in drawing no. 35 of the building's contract (1865).

(*above*): 5.15(b) Cast-iron column in the central octagon in the interior of the engine-house at Crossness (1862-65).

(*left*): 5.15(c) Cast-iron column supporting the platform canopy at Battersea Park railway station (1866), designed by Charles Driver.

(*centre*): 5.15(d) Cast-iron columns in the booking hall at Battersea Park railway station.

(*right*): 5.15(e) Charles Driver's extravagant cast-iron lamp, erected in 1868 and pictured in the *Builder*, 15 August 1868, p. 604.

market halls and glasshouses.[107] The structure supports both the packing floor – now removed, but originally consisting of a gallery on the intermediate level between the ground and first floor – and the beam floor above: a larger, galleried space with ornamental cast-iron railings throughout framing the openings for the beams of each of the eight engines. The internal octagon functions as both a structural support for the different levels of the building above and also as a showpiece for the ornamental possibilities of cast iron. To contemporaneous commentators, the octagon – enhanced by the raised stars in the arches, the massive columns and the Greek-cross plan of the building – strongly resembled the interior of a Byzantine church.[108]

Victorian interest in Byzantine architecture was, on the one hand, indicative of a general trend amongst architects in the 1860s towards stylistic eclecticism stimulated by an increasing knowledge of the history of art and, on the other, based on the particular associations of the Byzantine itself.[109] For Ruskin, the Byzantine style, seen particularly in Venice, represented both a transition from the Classical to the Gothic style[110] and also a fusion of western and eastern forms.[111] This popular notion of the Byzantine as a hybrid form – a synthesis of multifarious elements – would have appealed to architects like Driver who were searching for a way of accommodating the new forms generated by iron with traditional materials such as brick and stone. It also generated a breadth of interpretations of what exactly constituted the Byzantine, as is revealed most clearly in features of the Byzantine and Romanesque Court (**5.14: b**), designed by Matthew Digby Wyatt and Charles Fowler as one of a series of architectural courts in the Crystal Palace at Sydenham (1852-54). Wyatt's description of the Court emphasises the hybrid quality of the Byzantine,[112] justifying the bewildering array of forms in the Court drawn from Germany, Italy and Britain. From the mid-1850s onwards it was also one of the most accessible representations of the Byzantine, widely known amongst both public and professionals alike.

The octagonal arcaded structure in the centre of the interior of Abbey Mills corresponds closely to the Byzantine Court in both its appearance and in its eccentric fusion of different elements. The round arches, decorated with rope moulds and stars, the entablatures punctured by naturalistic ornament, and the massive, richly ornamented columns mirror the forms used in the Byzantine Court. The popular sense of the Byzantine as a transitional style is also reflected in Driver's opinions on iron: for him, iron was 'Byzantine' in that it suggested the possibility of a new style of architecture with the perceived transitional nature of the historical Byzantine forming a valid precedent. Consequently, iron was a material fit for 'noble' architectural treatment.

Columns

An examination of the columns that make up the most striking part of the octagonal structure (**5.15: a**) provides more specific evidence of Driver's ambitions with regard to cast iron. The columns are composed of distinct castings bolted together: the octagonal masonry base is linked to the cylindrical cast-iron shaft by means of an intervening subsidiary base of square-plan; the lower splayed part of the cylindrical shaft also continues this sense of transition; while the elaborate capitals are separately cast and bolted onto the shaft. Encouraged by Ruskin's promotion of the use of decorative spurs on column bases,[113] this method of smoothly linking different geometric forms was a popular device employed by contemporaneous architects like William Butterfield to imbue columns with an immediately apparent sense of structural strength and unity.[114] The unnecessary thickness of the columns at Abbey Mills (in terms of what is required for their structural function) also suggests an attempt to imbue iron

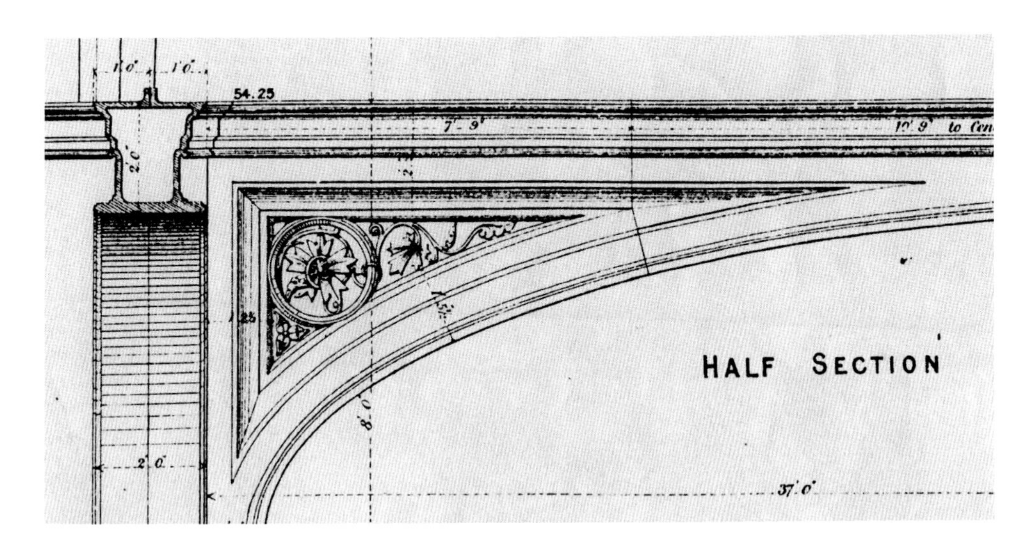

5.16(a) Cast-iron spandrel in a roof girder in the interior of Abbey Mills shown in drawing no. 37 of the building's contract (1865).

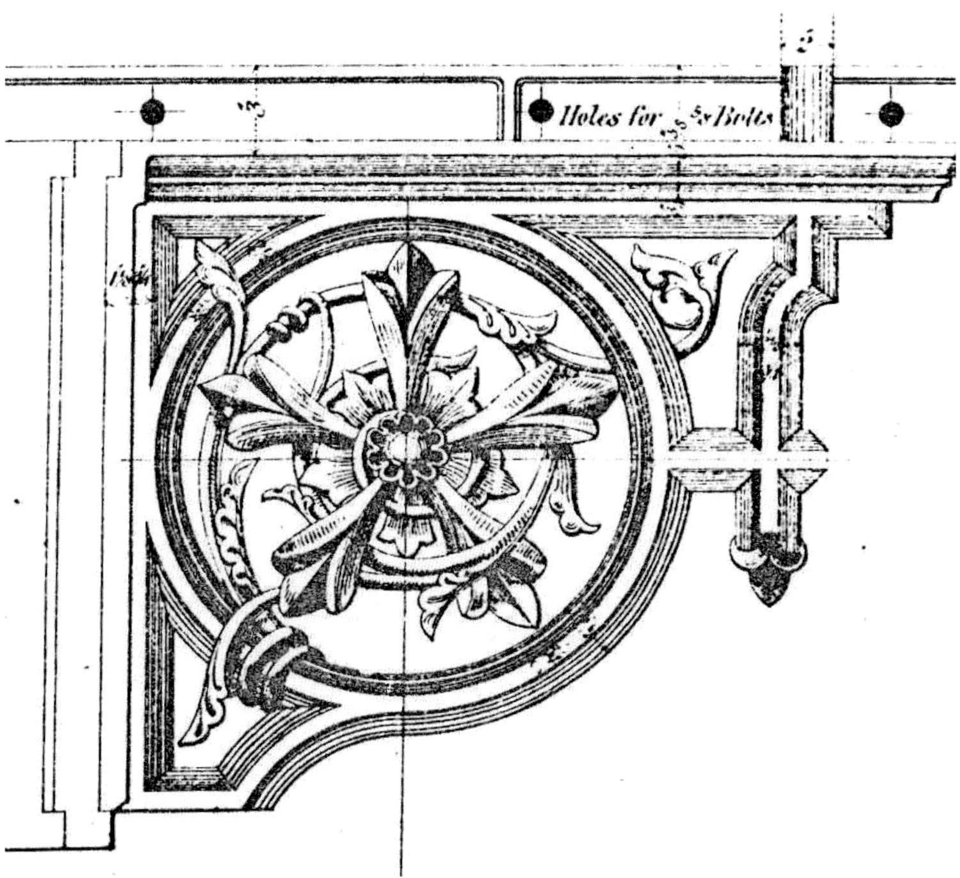

5.16(b) One of the many cast-iron brackets supporting the grille floor in the gallery of Abbey Mills and shown in drawing no. 37 of the building's contract (1865).

5.16(c) John Ruskin's proposed design for an wrought-iron spandrel in the interior court of the University Museum, Oxford (*c.*1855) and shown in John Acland and John Ruskin's *The Oxford Museum* (1859), p. 89.

5.16(d) Charles Driver's early use of cast-iron spandrels in the canopy platform of Wellingborough railway station, Northamptonshire (1857).

with a superadded sense of visible strength. These thick columns counteract Ruskin's objection to the inadequate sense of visible strength that he saw in the thin iron columns of the Crystal Palace.[115]

Further evidence is provided by the treatment of columns in Driver's other architectural projects. The cast-iron columns in the interior of the engine-house at Crossness (**5.15: b**) employ a similar pier-base as well as elaborate foliated capitals derived from the columns used at Victoria station (1860-61), the terminus building for the London, Brighton and South Coast Railway.[116] The cast-iron columns used by Driver in his stations for this railway show a similar highly individual treatment (**5.15: c**), with a complex variety of geometric forms in their shafts and bases. Those used in the booking hall at Battersea Park station (**5.15: d**) are arranged in an arched formation not dissimilar to that seen in the octagon in the interior of Abbey Mills, while their elaborately splayed bases and foliated capitals and marble-like paintwork also mirror the eccentric features of the individual columns in the octagon. The formal possibilities of iron columns was exploited most fully by Driver in his design for an elaborate gas lamp, erected in Holborn in August 1868 at the boundary between the Cities of Westminster and London (**5.15: e**).[117] With its octagonal cast-iron base – similar to a stone pier-base used by Butterfield in his restoration of Mapledurham church (1862-64) – and foliated caps rising to a central column of clustered shafts with a crested and foliated capital, this lamp lavishly displays the decorative potential of cast iron, drawing on Ruskinian notions of beauty and strength. The eccentric treatment of the columns in the interior of Abbey Mills corresponds closely with Driver's consistent attempts to produce new forms out of cast iron and to elevate its prosaic status by an ornamental treatment paradoxically derived from Ruskinian principles. Coupled with the articulation of similar concerns in his papers, such treatment confirms Driver's key role in the design of the interior of Abbey Mills.

5.17(a) Cast-iron spandrels supporting the roof of the train shed at London Bridge railway station (*c.*1866), designed by Charles Driver.

5.17(b) Driver's use of cast-iron spandrels in the entrance canopy at Leatherhead railway station.

5.17(c) Owen Jones's design for a shop front of Mr Chappell's house in Bond Street, London, 1850, in the *Journal of Design and Manufactures*, vol. 4 (1850), p. 13.

5.18(a) Cast-iron railing in the gallery space in the interior of Abbey Mills showing use of the motifs of roses, cones and leaves.

Spandrels, Brackets and Railings

Driver also exploits the decorative potential of iron in the spandrels and brackets seen throughout the internal space. All these are primarily structural features: the spandrels provide lateral support for the roof girders while the brackets support the beam gallery. The triangular space within the spandrel, however, is also used as an opportunity for decorative enrichment, in both the lower part of the roof structure (**5.16: a**) and in the supporting brackets (**5.16: b**). With their arboreal motifs and serpentine forms, the brackets and spandrels once again embody Ruskin's definitions of noble ornament. For Ruskin, all good ornamentation was 'arborescent', in that 'one class' of ornament should

5.18(b) Another element of the cast-iron railings at Abbey Mills: striking roundels, containing the coats of arms of the main sponsors of the main drainage system, surrounded by extravagant lilies.

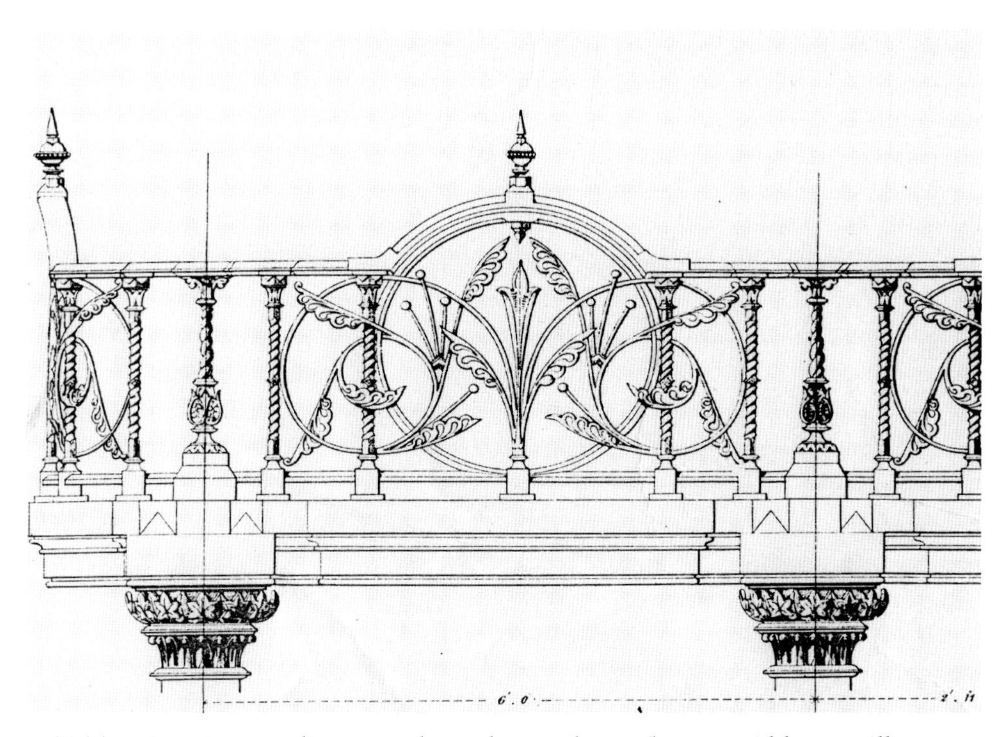

5.18(c) Cast-iron railings used in the packing floor at Abbey Mills, now demolished, but pictured in drawing no. 35 of the building's contract (1865).

'branch out of another' establishing a simple, obvious relationship of parts to the whole.[118] In his sole attempt to apply this theory to iron, Ruskin designed a spandrel for the interior of the Oxford Museum, with highly naturalistic horse chestnut leaves and nuts forming the principal motifs (**5.16: c**). The spandrel was acceptable to Ruskin only because it was made of wrought iron and was hand- not machine-made, and was therefore, according to his principles, capable of noble treatment. Unfortunately the spandrels, together with the rest of the ironwork, had to be modified due to their structural weakness.[119]

From the start of his career, Driver employed iron in spandrels in a very different way. Using only cast iron, his spandrels at Wellingborough station (1857; **5.16: d**), comprised of vine leaves set within a conventionalised arabesque form, show a much greater unity of form and function than Ruskin's equivalent attempt. The spandrels used by Driver in the London Bridge train shed (*c.*1866; **5.17: a**) are more geometric in character, while those at Leatherhead station (**5.17: b**) are highly individual and florid (and repeated in all of his stations for the South London Line). The profuse decoration of the spandrels and brackets at Abbey Mills, with both vine leaves and conventionalised flowers set in geometric serpentine forms, mirrors Driver's earlier treatment of these features. Such an approach followed Ruskin's dictum that all ornament should flow from a common source, a point reiterated by Driver himself in his 1875 paper to the R.I.B.A.[120] But Driver's approach also draws on similar advice offered by Owen Jones in *The Grammar of Ornament* (1856) and practically demonstrated in Jones's iron buildings of the 1850s, in particular his design for a shop front in 1850 (**5.17: c**).[121] Jones, like Ruskin, called for ornament that flowed out of a parent stem but, unlike Ruskin, argued that it should be set within a 'geometrical construction'.[122] Part of the reason for a conventionalised treatment of cast iron lay in its peculiar properties as a material and the logistics of the casting process; but another part relates to Driver's overarching synthesising aim: to bring together the respective poles of engineering and art.

Iron and Sewage

The interior of Abbey Mills displays perhaps the largest number of different cast-iron forms in any of Driver's projects, including – in addition to those already outlined above – repeated motifs of roses, cones and leaves; spiral columns with foliated bases and caps in the railings of the beam gallery (**5.18: a**); striking roundels, containing the coats of arms of the main sponsors of the main drainage system, surrounded by extravagant lilies (**5.18: b**); and the stiff-leaf forms of the original railings of the packing floor, now demolished (**5.18: c**). The sheer abundance of natural forms in the interior – whether in the octagon, iron columns, spandrels, brackets and railings – suggests the possibility that Driver might be positing a more direct correlation between sewage and natural abundance. In a lecture given to the Society of Mechanical and Civil Engineers on 19 December 1878, Driver summed up recent developments in engineering practice, drawing particular attention to the subjects of water supply and sanitation. Surprisingly, Driver calls into question the design of Bazalgette's main drainage system, which he viewed as making impossible the recycling of the London sewage as an agricultural fertiliser.[123] Enthusiasm for sewage utilisation was common in mid-Victorian Britain and was consistently put forward as a solution to the problem of human waste disposal, as discussed in more detail in chapter 2. By 1878, however, following the failure of many high-profile schemes that intended to recycle London's sewage, most enthusiasts had given up hope of ever achieving this.[124] The fact that Driver continued to assert the recycling imperative at this moment suggests that for him this was a long-held view. Consequently, the abundance of natural forms in the decoration of Abbey Mills, designed some ten years earlier, and their inclusion of agricultural as well as horticultural examples, may suggest a more direct and personal association between this abundant naturalism and the possible agricultural usefulness of the vast amounts of sewage that passed through the pumps every day. It might even be suggested that the interior of Abbey Mills proposes not only a new style for architecture, uniting the fragmentary disciplines of engineering and art, but also a new way of living for a new civilisation, based on the unification of man and his wastes.

Conclusion

The profusion of iron forms that Driver generated in the interior of Abbey Mills can be put into perspective through comparison with important examples designed by others in his field. Driver himself acknowledged, in 1875, that his approach to cast iron was very different from the influential precedents set by Matthew Digby Wyatt at Paddington Station and by Francis Skidmore in the interior of the University Museum, Oxford, both of which attempted to generate entirely new decorative forms out of iron.[125] Instead, Driver's predominantly historicist approach to cast iron reflects a change in taste in the 1860s, which now emphasised a synthesis of iron with more traditional building materials, such as brick and stone. Both Francis Fowke's buildings for the International Exhibition of 1862 and George Gilbert Scott's Midland Grand Hotel (1865-74) were important contemporaneous architectural projects that used iron alongside brick and stone, Fowke's interior ironwork also using similarly 'Byzantine' forms to those seen in the octagon at Abbey Mills.[126] The singularity of Driver's treatment of the ironwork at Abbey Mills lies not only in the sheer range of forms that he produces – far more than Fowke had attempted in 1862 – but also in the way in which his approach directly corresponds to that stated in his papers: a correspondence that cannot be said to exist between Scott's writing and practice in relation to iron. The ironwork at Abbey Mills demonstrates the extent to which Driver was aware of, and able to employ, the fullest range of historicist and hybrid forms then possible in cast iron, demonstrated most dramatically in the catalogues of iron manufacturers like Walter Macfarlane, who attended Driver's lecture in 1875 and worked closely with him on many of his projects concurrent with Abbey Mills.[127]

Such decorative extravagance stemmed from a fundamental problem in relation to Victorian architecture. For Driver, and many others in his field, utility possessed no meaning in itself, that is, lacked any aesthetic value required to make it 'architectural'.[128] For Driver, a synthesis – by contrast or otherwise – of utility and decoration was an important part of both his theorising and his architectural treatment of both traditional building materials and new ones, particularly iron. In the spectacular iron interior of Abbey Mills, Driver creates a hybrid or transitional iron style that synthesises a host of contrasting elements. Paradoxically, such hybridism also breaks down any sense of a oppositional relationship between decoration and utility (and consequently between the exterior and interior of the building), as proposed by Bazalgette and, later, so vehemently stressed by proponents of modernist architecture. In my discussion of the decorative elements of Abbey Mills, I stressed that all of the decoration features – the windows, doors, lantern, chimneys, decorative details, octagon, columns, spandrels, brackets and railings – possess both a utilitarian *and* symbolic function; form and function are fused together in a controlled, if eccentric, form of synthesis. In his contribution to Abbey Mills, Driver, acting as the self-defined 'true architect', clearly demonstrates that both contrast and synthesis were to be key elements in any modern treatment of building materials, particularly iron. The final chapter will assess the response to this sense of modernity when the pumping stations were used to present the main drainage system to those who were deemed to benefit from it and indeed those who would eventually pay for it: the citizens of London.

6

Conflated Spaces

In this final chapter, I return to the observations made at the start of the introduction to this book. John Hollingshead – the obsessive chronicler of London's sewers already introduced – summarised his extensive explorations of the city's underbelly with the statement: 'there are more ways than one of looking at sewers'.[1] Hollingshead may have delighted in the sewer as a contradictory space, at once a rational tool of the engineer and a monstrous repository of filth; yet even he might have been simply awestruck when faced with the lavish architectural display of the Crossness and Abbey Mills pumping stations at the opening ceremonies held in 1865 and 1868 respectively. In the event, Hollingshead was not invited to either ceremony but many of his fellow journalists were and, together, they left a voluminous record of responses that provides perhaps the best evidence of how the main drainage was perceived once it was completed.

The Public Role of the Pumping Stations

In contrast to the pumping stations at Deptford and Pimlico, Crossness and Abbey Mills were assigned the important role as sites for public ceremonies that marked the formal opening of the main drainage system. Despite it being unclear whether this 'public' role was a justification for their lavish architectural treatment, the striking appearance of Crossness and Abbey Mills suggests that these buildings had a 'civic' role more applicable to public buildings, such as town halls or churches. However, such a role was restricted to these ceremonies; in general, only the workmen, who tended the engines, would regularly see the lavish interior ironwork. Nevertheless, both Crossness and Abbey Mills were, and continue to be, important sites for the promotion of the main drainage system to the public – places where the vast but invisible sewerage system is 'summed up' in a celebratory aesthetic statement.[2]

The ceremonies held in 1865 at Crossness and 1868 at Abbey Mills marked the operational starting of the main drainage system and both were intended to be lavish events: the Prince of Wales being invited to Crossness and the Duke of Edinburgh to Abbey Mills, as well as many Members of Parliament and other important dignitaries.[3] In the event, Crossness was the more high profile event, due to the recess of Parliament and the unavailability of the Duke of Edinburgh in 1868.[4] Six hundred guests attended the ceremony at Crossness, which began at 11a.m. on 4 April 1865 with special trains laid on from Charing Cross to the remote site on the Essex marshes and a steamboat carrying the royal party from Westminster. The events of the day included tours of the underground sewage reservoir, an explanatory lecture by Bazalgette, a ceremony in the lavishly decorated engine-house (where the Prince of Wales started the engines), and a banquet in one of the workshops. The ceremony at Abbey Mills, on 31 July 1868, took place on the same day as the opening of the Victoria Embankment, a project concurrent and connected to the main drainage system, and it followed a similar, if stripped-down schedule to that at Crossness. Visits to Abbey Mills also continued after the main ceremony: during the following fortnight, representatives from London's vestries visited the pumping station in a succession of organised tours.

Both ceremonies followed the established precedent of holding public events to mark the completion of technological projects, especially in a subterranean context, where large parts were effectively invisible to the public. When the world's first tunnel under a river – the Thames Tunnel – was temporarily opened in 1827, the engineer Marc Brunel organised a banquet for the workers and distinguished guests inside the Tunnel itself; in December 1851, the *Illustrated London News* reported on a similar ceremony inside a vast new underground water reservoir at Croydon;[5] while on 17 January 1863, the same newspaper pictured a banquet held at Farringdon Street station to mark the opening of the Metropolitan Underground Railway.[6] The ceremonies at Crossness and Abbey Mills were similar events, designed to highlight, to dignitaries, sponsors and the press, the importance of such subterranean technological development and to give it a visible and striking form. The notable presence of the press at these ceremonies represented an important interface between those who conceived the project – Bazalgette, Driver, the contractors and the Metropolitan Board of Works – and those on whom it impacted, whether in social, economic or psychological terms.

London's Press

In the days following the ceremonies at Crossness and Abbey Mills, voluminous articles appeared in London's newspapers. In April 1865, most of the city's thirty-or-so daily and weekly newspapers drew directly for their articles on four accounts: in the *Standard*, the *Morning Post*, *The Times*, and the *Daily Telegraph*, with *The Times* forming the main source in 1868. Such obvious plagiarism was common practice in a highly competitive and burgeoning market for 'news'. Throughout the 1850s and 1860s, particularly after the repeal of stamp duty in 1855 and paper duty in 1861, London's press, until then dominated by *The Times*, witnessed a dramatic increase in competition as new, cheaper newspapers broadened their audience to the wider middle classes. By the mid-1860s, London's press consisted of three distinct types of publication: established and new daily newspapers, such as *The Times* and the *Daily Telegraph* (founded in 1855); weekly newspapers, including illustrated newspapers like

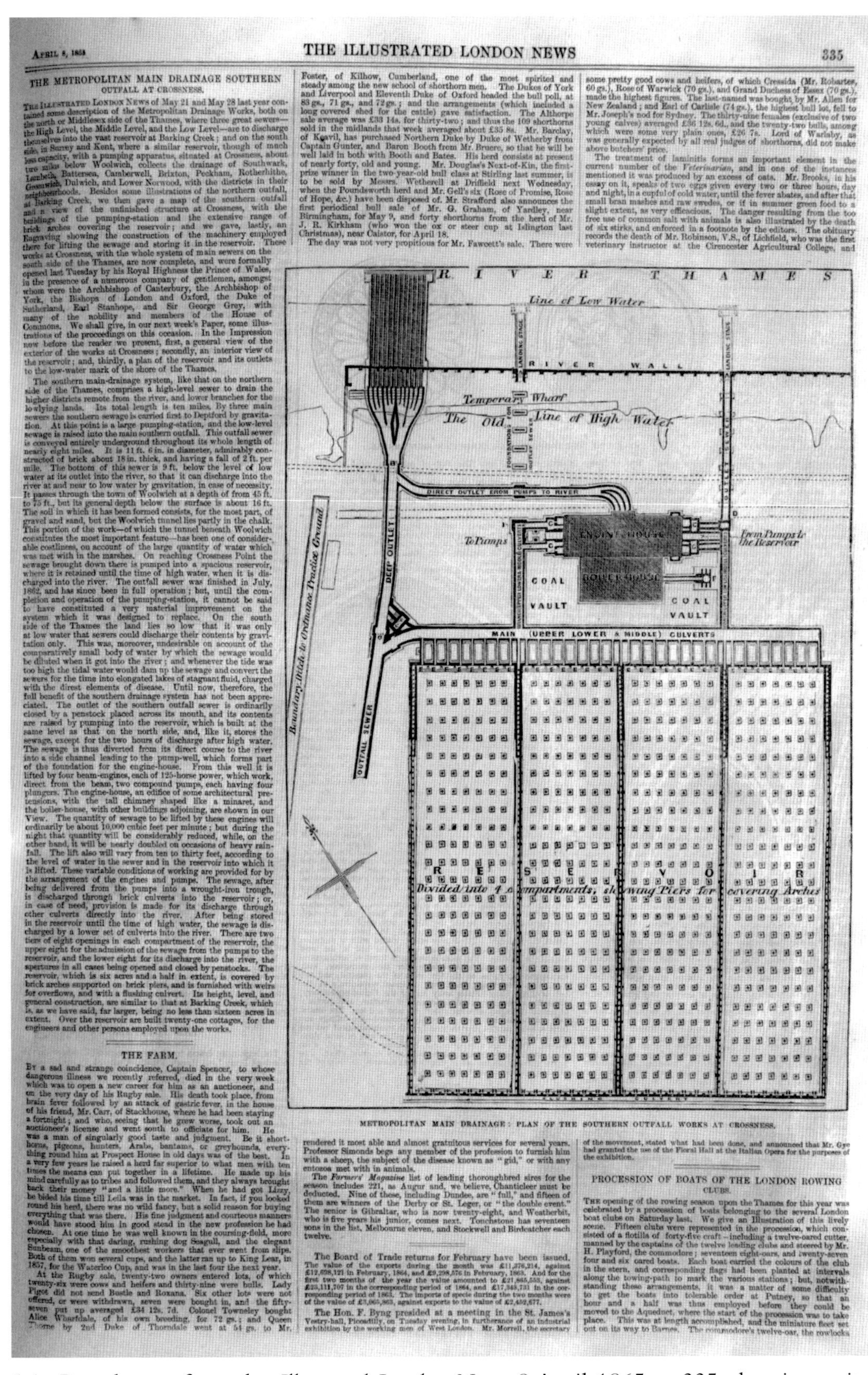

APRIL 8, 1865 THE ILLUSTRATED LONDON NEWS 335

THE METROPOLITAN MAIN DRAINAGE SOUTHERN OUTFALL AT CROSSNESS.

THE ILLUSTRATED LONDON NEWS of May 21 and May 28 last year contained some description of the Metropolitan Drainage Works, both on the north or Middlesex side of the Thames, where three great sewers—the High Level, the Middle Level, and the Low Level—are to discharge themselves into the vast reservoir at Barking Creek; and on the south side, in Surrey and Kent, where a similar reservoir, though of much less capacity, with a pumping apparatus, situated at Crossness, about two miles below Woolwich, collects the drainage of Southwark, Lambeth, Battersea, Camberwell, Brixton, Peckham, Rotherhithe, Greenwich, Dulwich, and Lower Norwood, with the districts in their neighbourhoods. Besides some illustrations of the northern outfall, at Barking Creek, we then gave a map of the southern outfall and a view of the unfinished structure at Crossness, with the buildings of the pumping-station and the extensive range of brick arches covering the reservoir; and we gave, lastly, an Engraving showing the construction of the machinery employed there for lifting the sewage and storing it in the reservoir. These works at Crossness, with the whole system of main sewers on the south side of the Thames, are now complete, and were formally opened last Tuesday by his Royal Highness the Prince of Wales, in the presence of a numerous company of gentlemen, amongst whom were the Archbishop of Canterbury, the Archbishop of York, the Bishops of London and Oxford, the Duke of Sutherland, Earl Stanhope, and Sir George Grey, with many of the nobility and members of the House of Commons. We shall give, in our next week's Paper, some illustrations of the proceedings on this occasion. In the Impression now before the reader we present, first, a general view of the exterior of the works at Crossness; secondly, an interior view of the reservoir; and, thirdly, a plan of the reservoir and its outlets to the low-water mark of the shore of the Thames.

The southern main-drainage system, like that on the northern side of the Thames, comprises a high-level sewer to drain the higher districts remote from the river, and lower branches for the lowlying lands. Its total length is ten miles. By three main sewers the southern sewage is carried first to Deptford by gravitation. At this point is a large pumping-station, and the low-level sewage is raised into the main southern outfall. This outfall sewer is conveyed entirely underground throughout its whole length of nearly eight miles. It is 11 ft. 6 in. in diameter, admirably constructed of brick about 18 in. thick, and having a fall of 2 ft. per mile. The bottom of this sewer is 9 ft. below the level of low water at its outlet into the river, so that it can discharge into the river at and near to low water by gravitation, in case of necessity. It passes through the town of Woolwich at a depth of from 45 ft. to 75 ft., but its general depth below the surface is about 16 ft. The soil in which it has been formed consists, for the most part, of gravel and sand, but the Woolwich tunnel lies partly in the chalk. This portion of the work—of which the tunnel beneath Woolwich constitutes the most important feature—has been one of considerable costliness, on account of the large quantity of water which was met with in the marshes. On reaching Crossness Point the sewage brought down there is pumped into a spacious reservoir, where it is retained until the time of high water, when it is discharged into the river. The outfall sewer was finished in July, 1862, and has since been in full operation; but, until the completion and operation of the pumping-station, it cannot be said to have constituted a very material improvement on the system which it was designed to replace. On the south side of the Thames the land lies so low that it was only at low water that sewers could discharge their contents by gravitation only. This was, moreover, undesirable on account of the comparatively small body of water by which the sewage would be diluted when it got into the river; and whenever the tide was too high the tidal water would dam up the sewage and convert the sewers for the time into elongated lakes of stagnant fluid, charged with the direst elements of disease. Until now, therefore, the full benefit of the southern drainage system has not been appreciated. The outlet of the southern outfall sewer is ordinarily closed by a penstock placed across its mouth, and its contents are raised by pumping into the reservoir, which is built at the same level as that on the north side, and, like it, stores the sewage, except for the two hours of discharge after high water. The sewage is thus diverted from its direct course to the river into a side channel leading to the pump-well, which forms part of the foundation for the engine-house. From this well it is lifted by four beam-engines, each of 125-horse power, which work, direct from the beam, two compound pumps, each having four plungers. The engine-house, an edifice of some architectural pretensions, with the tall chimney shaped like a minaret, and the boiler-house, with other buildings adjoining, are shown in our View. The quantity of sewage to be lifted by these engines will ordinarily be about 10,000 cubic feet per minute; but during the night that quantity will be considerably reduced, while, on the other hand, it will be nearly doubled on occasions of heavy rainfall. The lift also will vary from ten to thirty feet, according to the level of water in the sewer and in the reservoir into which it is lifted. These variable conditions of working are provided for by the arrangement of the engines and pumps. The sewage, after being delivered from the pumps into a wrought-iron trough, is discharged through brick culverts into the reservoir; or, in case of need, provision is made for its discharge through other culverts directly into the river. After being stored in the reservoir until the time of high water, the sewage is discharged by a lower set of culverts into the river. There are two tiers of eight openings in each compartment of the reservoir, the upper eight for the admission of the sewage from the pumps to the reservoir, and the lower eight for its discharge into the river, the apertures in all cases being opened and closed by penstocks. The reservoir, which is six acres and a half in extent, is covered by brick arches supported on brick piers, and is furnished with weirs for overflows, and with a flushing culvert. Its height, level, and general construction, are similar to that at Barking Creek, which is, as we have said, far larger, being no less than sixteen acres in extent. Over the reservoir are built twenty-one cottages, for the engineers and other persons employed upon the works.

THE FARM.

BY a sad and strange coincidence, Captain Spencer, to whose dangerous illness we recently referred, died in the very week which was to open a new career for him as an auctioneer, and on the very day of his Rugby sale. His death took place, from brain fever followed by an attack of gastric fever, in the house of his friend, Mr. Carr, of Stackhouse, where he had been staying a fortnight; and who, seeing that he grew worse, took out an auctioneer's license and went south to officiate for him. He was a man of singularly good taste and judgment. Be it shorthorns, pigeons, hunters, Arabs, bantams, or greyhounds, everything round him at Prospect House in old days was of the best. In a very few years he raised a herd far superior to what men with ten times the means can put together in a lifetime. He made up his mind carefully as to tribes and followed them, and they always brought back their money "and a little more." When he had got Lizzy, he bided his time till Leila was in the market. In fact, if you looked round his herd, there was no wild fancy, but a solid reason for buying everything that was there. His fine judgment and courteous manners would have stood him in good stead in the new profession he had chosen. At one time he was well known in the coursing-field, more especially with that daring, rushing dog Seagull, and the elegant Sunbeam, one of the smoothest workers that ever went from slips. Both of them won several cups, and the latter ran up to King Lear, in 1857, for the Waterloo Cup, and was in the last four the next year.

At the Rugby sale, twenty-two owners entered lots, of which twenty-six were cows and heifers and thirty-nine were bulls. Lady Pigot did not send Bustle and Roxana. Six other lots were not offered, or were withdrawn, seven were bought in, and the fifty-seven put up averaged £34 12s. 7d. Colonel Towneley bought Alice Wharfdale, of his own breeding, for 72 gs.; and Queen Thorne by 2nd Duke of Thorndale went at 54 gs. to Mr. Foster, of Kilhow, Cumberland, one of the most spirited and steady among the new school of shorthorn men. The Dukes of York and Liverpool and Eleventh Duke of Oxford headed the bull poll, at 83 gs., 71 gs., and 72 gs.; and the arrangements (which included a long covered shed for the cattle) gave satisfaction. The Althorpe sale average was £33 14s. for thirty-two; and thus the 109 shorthorns sold in the midlands that week averaged about £35 8s. Mr. Barclay, of Keavil, has purchased Northern Duke by Duke of Wetherby from Captain Gunter, and Baron Booth from Mr. Bruere, so that he will be well laid in both with Booth and Bates. His herd consists at present of nearly forty, old and young. Mr. Douglas's Next-of-Kin, the first-prize winner in the two-year-old bull class at Stirling last summer, is to be sold by Messrs. Wetherell at Driffield next Wednesday, when the Poundsworth herd and Mr. Gell's six (Rose of Promise, Rose of Hope, &c.) have been disposed of. Mr. Strafford also announces the first periodical bull sale of Mr. G. Graham, of Yardley, near Birmingham, for May 9, and forty shorthorns from the herd of Mr. J. R. Kirkham (who won the ox or steer cup at Islington last Christmas), near Caistor, for April 18.

The day was not very propitious for Mr. Fawcett's sale. There were some pretty good cows and heifers, of which Cressida (Mr. Robartes, 60 gs.), Rose of Warwick (70 gs.), and Grand Duchess of Essex (70 gs.), made the highest figures. The last-named was bought by Mr. Allen for New Zealand; and Earl of Carlisle (74 gs.), the highest bull lot, fell to Mr. Joseph's nod for Sydney. The thirty-nine females (exclusive of two young calves) averaged £36 12s. 6d., and the twenty-two bulls, among which were some very plain ones, £26 7s. Lord of Warlaby, as was generally expected by all real judges of shorthorns, did not make above butchers' price.

The treatment of laminitis forms an important element in the current number of the *Veterinarian*, and in one of the instances mentioned it was produced by an excess of oats. Mr. Brooks, in his essay on it, speaks of two eggs given every two or three hours, day and night, in a cupful of cold water, until the fever abates, and after that small bran mashes and raw swedes, or if in summer green food to a slight extent, as very efficacious. The danger resulting from the too free use of common salt with animals is also illustrated by the death of six stirks, and enforced in a footnote by the editors. The obituary records the death of Mr. Robinson, V.S., of Lichfield, who was the first veterinary instructor at the Cirencester Agricultural College, and rendered it most able and almost gratuitous services for several years. Professor Simonds begs any member of the profession to furnish him with a sheep, the subject of the disease known as "gid," or with any entozoa met with in animals.

The *Farmers' Magazine* list of leading thoroughbred sires for the season includes 221, as Augur and, we believe, Chanticleer must be deducted. Nine of these, including Dundee, are "full," and fifteen of them are winners of the Derby or St. Leger, or "the double event." The senior is Gibraltar, who is now twenty-eight, and Weatherbit, who is five years his junior, comes next. Touchstone has seventeen sons in the list, Melbourne eleven, and Stockwell and Birdcatcher each twelve.

The Board of Trade returns for February have been issued. The value of the exports during the month was £11,376,214, against £12,898,121 in February, 1864, and £9,298,576 in February, 1863. And for the first two months of the year the value amounted to £21,865,555, against £23,111,707 in the corresponding period of 1864, and £17,349,731 in the corresponding period of 1863. The imports of specie during the two months were of the value of £3,065,863, against exports to the value of £2,452,677.

The Hon. F. Byng presided at a meeting in the St. James's Vestry-hall, Piccadilly, on Tuesday evening, in furtherance of an industrial exhibition by the working men of West London. Mr. Morrell, the secretary of the movement, stated what had been done, and announced that Mr. Gye had granted the use of the Floral Hall at the Italian Opera for the purposes of the exhibition.

METROPOLITAN MAIN DRAINAGE: PLAN OF THE SOUTHERN OUTFALL WORKS AT CROSSNESS.

PROCESSION OF BOATS OF THE LONDON ROWING CLUBS.

THE opening of the rowing season upon the Thames for this year was celebrated by a procession of boats belonging to the several London boat clubs on Saturday last. We give an Illustration of this lively scene. Fifteen clubs were represented in the procession, which consisted of a flotilla of forty-five craft—including a twelve-oared cutter, manned by the captains of the twelve leading clubs and steered by Mr. H. Playford, the commodore; seventeen eight-oars, and twenty-seven four and six oared boats. Each boat carried the colours of the club in the stern, and corresponding flags had been planted at intervals along the towing-path to mark the various stations; but, notwithstanding these arrangements, it was a matter of some difficulty to get the boats into tolerable order at Putney, so that an hour and a half was thus employed before they could be moved to the Aqueduct, where the start of the procession was to take place. This was at length accomplished, and the miniature fleet set out on its way to Barnes. The commodore's twelve-oar, the rowlocks

6.1 Page layout from the *Illustrated London News*, 8 April 1865, p. 335, showing a site plan of the Crossness pumping station, the outfall and the sewage reservoir.

6.2 A general view of the extensive site of the Crossness pumping station as pictured in the *Illustrated London News*, 8 April 1865, p. 325, which would eventually include workshops, workers' houses and a school.

the *Illustrated London News*; and specialist journals, like the *Builder*, published weekly, fortnightly or monthly. The reduction in prices – many to one penny – after the tax repeals stimulated intense competition, especially amongst the daily newspapers, where the dominance of *The Times*, which maintained its price at three pence after 1861, began to be challenged. By the mid-1860s, four main 'dailies' took the largest share of the market: *The Times*, with approximately 50,000 readers every day; the *Standard* with 60-70,000; and the *Daily Telegraph* and *Daily News* with upwards of 100,000 each. In terms of their ideological

stance, both *The Times* and *Standard* were largely conservative and appealed to a more 'respectable' middle-class audience, while the *Daily Telegraph* articulated a more radical agenda with 'vigorous and versatile writing' that had a broader appeal, with the *Daily News* falling somewhere in between.[7]

In the context of responses to the ceremonies at Crossness and Abbey Mills, accounts in *The Times* and the *Standard* formed the main source for other press accounts, with the *Morning Post* representing a significant minority wanting a more 'aristocratic' tone to their news;[8] accounts in the *Daily Telegraph*, the other significant source, lived up to the newspaper's reputation with their florid language and poetic embellishments, as will be seen below. It is important to stress at the outset the key role of these 'source' accounts, which formed the basis not only for countless articles in other newspapers, but also for the text and engravings in the *Illustrated London News*, whose editors, writers and

328 THE ILLUSTRATED LONDON NEWS April 8, 1865

THE METROPOLITAN MAIN-DRAINAGE WORKS AT CROSSNESS: VIEW IN THE RESERVOIR.—SEE PAGE 335.

hind part, from the breech to the trunnions, a length of about six feet, is protected, in the first place, by a steel jacket four inches and a half thick forced on over the inner tube, and fitting at each end to certain projections in the central piece, at the breech and at the trunnions respectively, by which the jacket is locked up in its place; the end of it, moreover, being so formed as to give support to the rear of the gun, so that it serves at once to resist the longitudinal and the lateral pressure. Above this jacket are two layers which have a thickness of six inches in that part of the gun behind the trunnions, but are slighter in the front part, where the strain is inconsiderable, and are finally rounded off between the trunnions and the muzzle, as appears in our Illustration. These pieces are so placed that the junction of those beneath is overlapped by the pieces of the outer layer. All the gun is of steel, except the trunnions, which are of Low Moor iron. The terminal knob, or cascable, fitted into the rear of the gun, is independent of the twelve inches solid thickness provided for in the central tube. Such is the gun itself. We have now to describe how it is mounted and furnished for working. It may be said to have a double carriage, or, rather, a carriage which travels upon a traversing platform or slide. The platform below turning on a central pivot, and on the semicircular railway behind, enables the gun to be trained, so as to point its muzzle right or left. The carriage above runs down and up the inclined plane of the platform, as the gun is loaded and discharged. With reference to the lower, or platform movement, we should observe that the whole weight of gun and carriage, amounting to thirty-four tons, can be moved upon the rail by a single man working the gear from behind; whereas, in the case of other great guns travelling upon a similar railway, it has been necessary, for want of such gearing, to train the gun either by means of handspikes or tackle requiring a number of men to work them. The contrivances for the upper, or carriage, movement are no less worthy of remark. Instead of making the hind part of the carriage rest upon blocks of timber, to stop the recoil of the gun—a method which necessitates the partial lifting of the gun, with levers or handspikes, whenever it has to be run out forward—the whole carriage here rests upon four rollers of cast iron; and it runs freely, by gravitation, down the inclined plane of the platform, or runs back with the recoil when discharged. This movement, for a length of from 7ft. to 10ft. up and down the platform, is controlled with perfect ease and precision by the use of the compressor, a sort of brake, which is worked by a man at the capstan wheel attached to the side of the carriage. This compressor itself is formed by a series of longitudinal plates of iron, which are suspended under the bottom of the carriage and lie in the intervals of a series of wooden bars which lie parallel to them, running the whole length of the platform, and, while in this position, they are grasped by a pair of very powerful levers, holding the entire series of wooden and iron bars together, so that the motion of the carriage along the platform may either be retarded by producing a moderate degree of friction, or else it may be quickly checked. As soon as the compressor is loosened the carriage runs forward by its own weight; but after firing, it runs back with the recoil, and is then held by a compressor till reloaded. The apparatus for loading is conspicuous enough in our Illustration, which shows a sort of crane, overhanging the muzzle of the gun, with a pulley and chain or rope worked by a windlass; at the end of the rope are a pair of Lewis (double-hinged) tongs, in which an object, when once inclosed by the claws of the forceps, is held by its own weight. The shot is lifted by this means to the level of the muzzle, where it is met by the shot-guide, a half-hoop or half-cylinder, which descends upon its hinges to catch the shot and hold it, as in the hollow of a spoon, while the men put it into the gun and ram it home in the ordinary manner. We are informed, however, that the Blakely Company have devised a further improvement, which will supersede the overhanging crane; the shot being raised to the muzzle, and held there, by the iron arms of a machine with a similar action to that of a man's hands bringing his food to his mouth, and which (except when in use) will not project above the gun. Our Illustration further shows

THE BLAKELY GUN MANUFACTURED FOR THE RUSSIAN GOVERNMENT.

6.3 Page layout from the *Illustrated London News*, 8 April 1865, p. 328, showing the interior of the sewage reservoir at Crossness (upper engraving).

artists would have gleaned descriptive details from these source accounts. Consequently, that which most widely represents a 'press' response is embodied in these key source articles.

Rational Spaces

Large sections of the press articles describing both ceremonies were effectively technical accounts of the main drainage system and the pumping stations, drawn from descriptions by Bazalgette. During the ceremony at Crossness in 1865, Bazalgette gave a lecture in one of the workshops, adapted from another talk he gave at the Institution of Civil Engineers in March 1865.[9] On 4 April 1865, articles in *The Times*, *Standard*, and *Morning Post* included long extracts from this lecture, mostly in the form of a series of precise but impressive facts and figures, such as the 82 miles of new sewers, the 318 million bricks and 880,000 cubic yards of concrete used, or the three and a half million cubic yards of earth excavated.[10]

When the *Illustrated London News* published its own account of Crossness on 8 April 1865, it also included a long technical description of the site as well as three wood-engraved illustrations: a plan of the site (**6.1**), a general view (**6.2**), and a view of the interior of the subterranean reservoir (**6.3**). In its article, the newspaper states to its readers that it would be illustrating the ceremony in its next issue, as the engravings were still in the process of being prepared.[11] The three engravings in this issue relate very closely to the technical clarity of the article, which corresponds with Bazalgette's lecture and accounts published by the *Standard, The Times* and *Morning Post*. Indeed, the ground plan of the site (**6.1**), embedded within the text of the article, is actually a modified version of one of Bazalgette's contract drawings showing the precise functional arrangement of the site and its various components: the engine-house, boiler-house, the outfall sewers, and the giant sewage reservoir.[12] In setting this precise engineering drawing within the text that describes it, the page layout suggests an immediate correspondence between text and image, enhancing the educative and technical role of both. The general view (**6.2**) is a three-dimensional version of the plan with picturesque additions, such as the dramatic sky and figures in the foreground, presumably visitors being transported to the site. The image is rendered along the same axis as the plan so that the two images can be easily read together, in order to give the reader/viewer a more comprehensive educational picture. The engraving of the interior of the reservoir (**6.3**), although much darker than its counterparts, is nevertheless an image that brings out technical, rather than dramatic, aspects of its spaces. Like **6.2**, this engraving also complements and expands upon the plan view (**6.1**): it emphasises the precise forms of the brick arches and concrete piers and also shows, prominent in the right foreground, one of the penstocks, or gates, that separated the four compartments of the reservoir. All three engravings relate closely both to the text and also to each other with **6.2** and **6.3** giving a comprehensible visual form to the technical details given in both the text and the plan view.

After the ceremony at Abbey Mills in 1868, such technical details made up the bulk of the press accounts; these were drawn more directly from a descriptive account of the building written by Bazalgette especially for the occasion and printed and distributed to all the visitors.[13] Much of the long article published in *The Times* after the ceremony on 31 July 1868 was directly copied from Bazalgette's account; this article formed the basis for most of the other press coverage of the event.[14] Bazalgette's description of Abbey Mills focused on the building's qualities as an engineering achievement and, despite precisely describing its architectural details, gives no suggestion as to any symbolic meaning or aesthetic considerations. Rather, it reinforced, to those who visited its spaces, the notion of Abbey Mills as a rational, functional building, precisely tailored to fulfil its engineering duty. In relation to the accounts of the Crossness ceremony, which were only partly informed by Bazalgette's own descriptions of the site, press responses to Abbey Mills were much more in line with the engineer's viewpoint. Certainly, there seems to have been a more direct intention on the part of Bazalgette to inform the press as to the rationalistic principles underlying his system and, as a consequence, to direct attention away from the aesthetic impact of the building.

When the *Illustrated London News* published its article describing the ceremony on 15 August 1868, its long description of Abbey Mills was drawn almost entirely from Bazalgette's account.[15] It also included two engravings of the building, arranged on one page: one showing a general view of the engine-house and the other, directly below, picturing the interior (**5.1**). Whilst these engravings highlight the extravagance of the design, both inside and out, their dramatic visual impact is offset by the technical and prosaic tone of the accompanying article on the adjacent page. Indeed, such exterior/interior image combinations were commonly employed by the newspaper to comprehensively depict a particular scene or event in order to educate its readers/viewers. This 'documentary' role of the images both reflected the newspaper's attitude towards wood engraving as a medium suitable for technical exposition and Bazalgette's attitude towards Abbey Mills, expressed in his rationalistic description of the building.

Magical Spaces

However, such rational description cannot be considered in isolation. Alongside, and often because of such facts and figures, some press accounts of the ceremonies relate a sense of the magical quality of the main drainage system. The 'extraordinary statistics'[16] provided by Bazalgette led some journalists, especially in 1865, to compare the new sewers with the wonders of the ancient world.[17] According to the *Daily Telegraph*, the main drainage system was a project alongside which even the Pyramids of Egypt and the sewers of Rome 'paled into comparison'.[18] In 1868, the *Marylebone Mercury* made similar comparisons: the main drainage system is described as the 'representation of a mighty civilisation' – a civilisation nobler than ancient Rome because it lacked its 'despotic power'.[19] Such comparisons transform statistics into myth: the impressive facts and figures provoke wonder at what many saw as a monument to the future when London, especially compared with its main rival, Paris, would become the cleanest and most magnificent city the world had ever seen.[20] This mythmaking replicates that seen in press responses to the tours of the construction sites in 1861 (see chapter 4); the articles in 1865 and 1868 mirror their tone of rhetorical sublimity. If the content of London's new sewers was 'not a bit better' than that in the sewers of Paris,[21] their technological and political basis most certainly was. Under Napoleon III and Baron Haussman, Paris was undergoing a more brutal transformation than London, with wide boulevards being driven through the medieval city, the new sewers constructed beneath them. Whilst some criticised the Metropolitan Board of Works for not 'Haussmanizing' London enough,[22] most celebrated the city's new sewers as making the city 'above comparison with any other European capital'.[23] If Bazalgette had done 'what Tarquin did for Rome' he did it without the 'despotic power' of the latter.[24] The fact that Napoleon III was also self-consciously modelling his new Paris sewers on this Roman precedent points to an implicit criticism of his 'despotic' methods as well.[25] According to the *Observer*, at Crossness, unlike in Paris, 'all shades of politics' came together to celebrate 'this great national work'.[26]

In relation to the events at Crossness in 1865, there were two aspects that brought out its magical quality most insistently: the interior of the engine-house and the subterranean sewage reservoir. If, according to the *Standard*, an 'enchanter's wand' had touched the whole site at Crossness, the interior of the engine-house – with its elaborate, brightly-painted decorative ironwork and giant steam engines – is described as a 'perfect shrine of machinery'.[27] According to the *Daily News*, the 'beautiful octagon' in the centre of the engine-house resembled the interior of a Byzantine church, with the shafts of the steam engines acting as 'church galleries – the pulpit being supplied by the cylinder'.[28] Press accounts of the Abbey Mills engine-house lacked such direct religious associations, but some of the articles did refer to the 'tremendous engines',[29] the 'wonderful machinery'[30] and to a 'deep wonder and admiration' at the sight of the lavish decorative ironwork.[31] The sense in which, according to the *Daily Telegraph*, the 'factory becomes poetical' and the 'furnace, fairy-like'[32] strongly relates to the perceived reconciliation of the artistic and the useful in these spaces; put another way, the imbuing of the purely functional with symbolism normally reserved for religious buildings made the prosaic seem magical.

Religious associations were also made during the visit to Crossness's vast underground sewage reservoir, where one of its four compartments had been kept free of sewage for the visitors. Compared to the gigantic crypt of a gothic cathedral,[33] it was lit especially for the occasion with, according to the *Daily Telegraph*, '100,000 coloured lamps, which produced a fairy-like appearance'.[34] Some compared the effect to that experienced at night in London's pleasure gardens at Cremorne and Vauxhall;[35] all were astonished and pleased by its striking appearance – the *Daily Telegraph* stating that it was 'bewildering in its beauty' and comparing it, oddly enough, to the 'piazza of St Mark's by night'.[36] This was truly a 'subterranean wonder' and was the object of greatest interest to the visitors. Press accounts of the ceremony at Abbey Mills three years later lacked such dramatic associations, perhaps because there was no comparable subterranean 'wonder'; the tone of most of the articles was prosaic and explanatory, like Bazalgette's account that formed their source.

When the *Illustrated London News* published its second account of Crossness on 15 April 1865, it included a further five large-scale wood engravings depicting the ceremony in the engine-house (**6.4**), the interior ironwork of the engine-house, Bazalgette's lecture and the banquet in one of the workshops (**6.5**), and the subterranean sewage reservoir (**6.6**).[37] The text of the article on page 342 is largely borrowed from the earlier account in the *Morning Post* and its tone is both celebratory and prosaic. However, the engravings, especially when compared to those published a week earlier by the newspaper, highlight, in visual terms, the magical quality perceived in some of the press accounts already discussed. Compared with the technical plan published on 8 April (**6.1**), the engraving showing the ceremony in the engine-house (**6.4**) represents both architectural detail and a theatrical 'event' acted out within its spaces: that is, the Prince of Wales – shown in the centre-right of the image and surrounded by a crowd of cheering figures – pulling a lever to start the engines. The cylinder of one of the engines – seen in the centre of the engraving with three figures standing on top of it – does indeed resemble a church 'pulpit', as the *Daily News* had remarked on 5 April, with the 'church galleries' behind – in reality the supporting floors of the steam engines – also crowded with onlookers. Shown in the extreme right of the engraving is part of the 'beautiful octagon', described by the *Daily News*, with its foliated cast-iron capitals and entablatures; while on the left, in the arch above the window, is revealed some of the 'handsome' brickwork praised in many of the press accounts. The engraving gives a visual form to what the non-illustrated press accounts had stressed: that the perceived synthesis of architecture and engineering (or the artistic and the useful) in the engine-house made the functional take on religious or magical associations. Furthermore, as a front-page image, like that showing the bursting of the Fleet sewer in 1862 (**4.10**), this engraving functions as a dramatic visual introduction to this particular issue of the newspaper. As such, it reflects not only the dramatic language of the non-illustrated accounts but also the versatility of wood engraving as a medium: on the one hand, wood engraving could represent the sewer system as a technical achievement with engravings copied directly from Bazalgette's engineering drawings; on the other, it could picture the magical quality of the spaces inside the engine-house.

In its 15 April issue, the *Illustrated London News* also included a full-page stand-alone engraving showing the interior of the sewage reservoir on the day of the ceremony, lit up by the '100,000 coloured lamps' described by the *Daily Telegraph* (**6.6**). Compared with the engraving showing a similar view a week earlier (**6.3**), this image, like that on the front page (**6.4**), also gives a striking visual form to the transformed perception of these spaces, described in the non-illustrated press accounts. This engraving is much larger than its counterpart, filling an entire page and separated from the article that describes it by six pages, further accentuating its dramatic impact as a stand-alone image. The viewpoint, positioned in the very centre of the reservoir, stresses the dramatic – seemingly

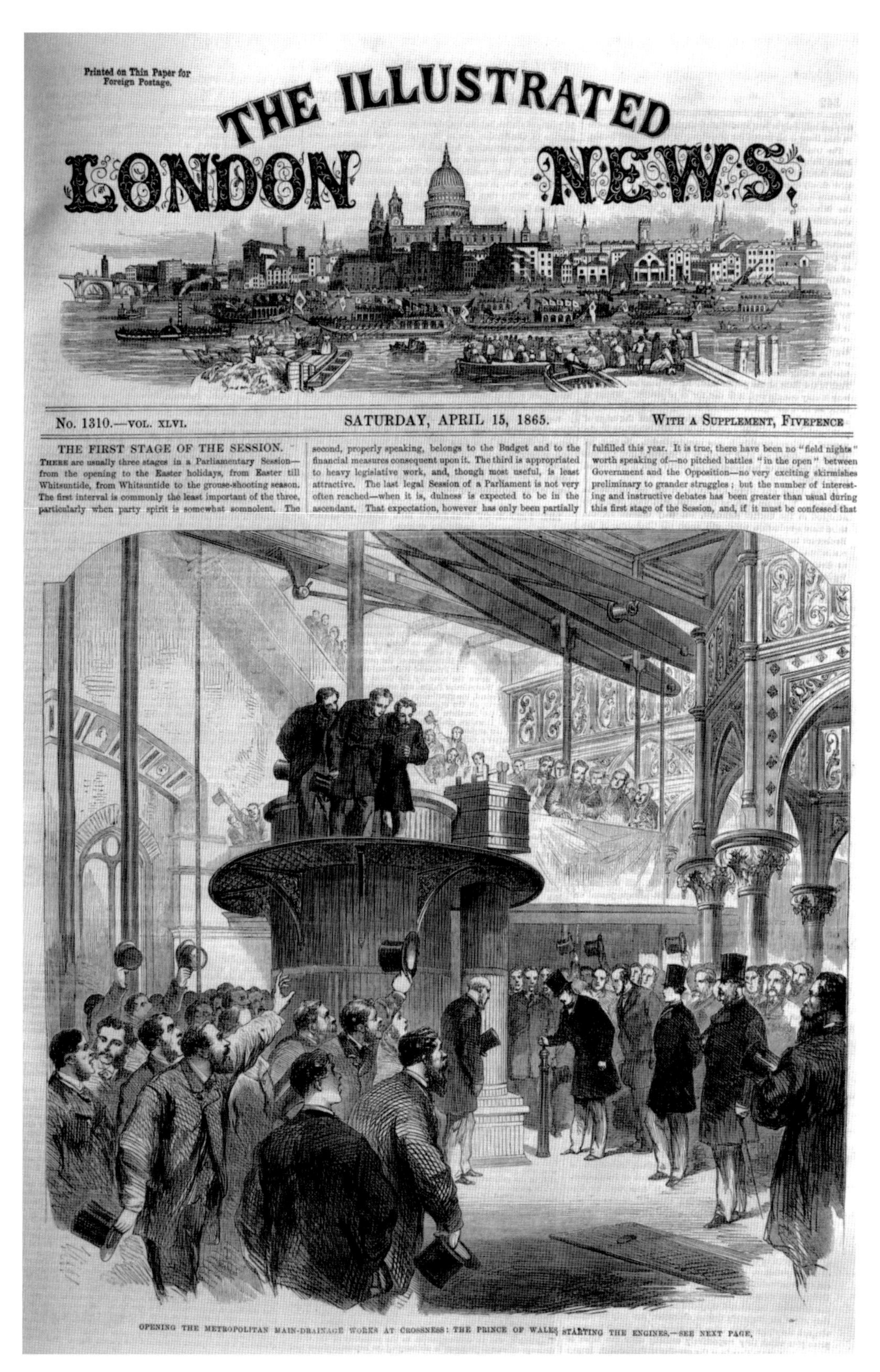

Printed on Thin Paper for Foreign Postage.

THE ILLUSTRATED LONDON NEWS.

No. 1310.—VOL. XLVI. SATURDAY, APRIL 15, 1865. WITH A SUPPLEMENT, FIVEPENCE

THE FIRST STAGE OF THE SESSION.

THERE are usually three stages in a Parliamentary Session—from the opening to the Easter holidays, from Easter till Whitsuntide, from Whitsuntide to the grouse-shooting season. The first interval is commonly the least important of the three, particularly when party spirit is somewhat somnolent. The second, properly speaking, belongs to the Budget and to the financial measures consequent upon it. The third is appropriated to heavy legislative work, and, though most useful, is least attractive. The last legal Session of a Parliament is not very often reached—when it is, dulness is expected to be in the ascendant. That expectation, however has only been partially fulfilled this year. It is true, there have been no "field nights" worth speaking of—no pitched battles "in the open" between Government and the Opposition—no very exciting skirmishes preliminary to grander struggles; but the number of interesting and instructive debates has been greater than usual during this first stage of the Session, and, if it must be confessed that

OPENING THE METROPOLITAN MAIN-DRAINAGE WORKS AT CROSSNESS: THE PRINCE OF WALES STARTING THE ENGINES.—SEE NEXT PAGE.

6.4 The Prince of Wales turns on the steam engines in the interior of the engine-house at Crossness, as seen on the front page of the *Illustrated London News*, 15 April 1865.

344 THE ILLUSTRATED LONDON NEWS APRIL 15, 1865

THE PRINCE OF WALES OPENING THE METROPOLITAN MAIN-DRAINAGE WORKS AT CROSSNESS.

MR. BAZALGETTE EXPLAINING THE MAIN-DRAINAGE PLANS.

THE LUNCHEON IN THE WORKSHOP AT CROSSNESS.—SEE PAGE

6.5 Page layout from the *Illustrated London News,* 15 April 1865, p. 344, showing Bazalgette giving a lecture on the main drainage system (upper engraving) and the banquet in one of the workshops adjoining the engine-house (lower engraving).

infinite – recess of the arches lit up by the myriad lamps, while the prominent figures in the foreground further accentuate the vast scale of the enclosed space. Unlike **6.3**, which concentrates on the technical aspects of the reservoir, such as the penstocks separating the compartments, this engraving focuses on the dramatic qualities of the space itself. By emphasising the repeating arches, their dramatic recession into space, and the contrasts of light and dark caused by the lamps, this engraving gives a visual form to the sense of magical sublimity perceived by the press in the accounts discussed previously.

Monstrous Spaces

Alongside the perceived magical quality of the underground reservoir at Crossness was a more disturbing sense of the monstrous. The writer for the *City Press*, describing the descent from the 'warm daylight' into 'strange dimly defined vistas', confesses to a 'curious nervous shock (not disagreeable)'.[38] The appearance of the reservoir – brilliantly lit and empty of sewage – led the *Morning Star* to state that this was not what you would naturally expect in such a place – that is, the 'foul, filthy and abominably nasty'.[39] However, it is the writer for the *Daily Telegraph*, of the most 'poetic turn of mind', who plays most strongly on this disassociation of imagination and reality. If the reservoir was so clean 'you could have eaten your dinner off it' it concealed, in the parts already filled with sewage, 'a repulsive flood'. If there was 'no foul festona or feculent moisture' in this part of the reservoir, then 'light would soon give way to darkness, dirt [and] rats' when the visitors left and the reservoir was filled with sewage and 'shut away from the public gaze forever'.[40]

Indeed, this writer revelled in the unique conjunction of the clean and the dirty: standing in the empty part of the reservoir, with its fairy lights and crypt-like space, the close proximity of the sewage in other unseen parts of the reservoir prompted the writer to feel 'in the very jaws of peril, in the gorge of the valley of the shadow of death', separated only by bolted iron gates from the 'the filthiest mess in Europe, pent up and bridled in, panting and ready to leap out like a black panther at the turning of a wheel, at the loosening of a trap, at the drawing of a bolt'.[41] Why was this writer, in particular, so affected by the space in the reservoir? Certainly, in general, the language of the *Daily Telegraph* tended to be more 'vigorous' than its main 'conservative' rivals, *The Times* and the *Standard*. Nevertheless, compared to other articles in the *Daily Telegraph* describing similar events, such as the opening of the Metropolitan Underground Railway in January 1863, this account of the Crossness reservoir is singularly extravagant in its poetic excesses.[42] There seems to have be a unique quality of this space that stimulated an imaginative response on the part of the press, given strongest expression by this writer.

Another reason for the appeal of this subterranean space was the fact that ladies were excluded. From the *Illustrated London News*'s depiction of the banquet it is clear that women were present at the ceremony – but even in this 'clean' space they were physically raised above the ground and separated from the crowd of men below (**6.5**). The sewage reservoir, despite not containing its foul 'tenant', was no place for a lady and the language of the press accounts is

6.6 Male visitors being instructed by Bazalgette inside the subterranean sewage reservoir at Crossness, as seen in the *Illustrated London News*, 15 April 1865, p. 348.

348
THE ILLUSTRATED LONDON NEWS
APRIL 15, 1865
OSSNESS: THE UNDERGROUND RESERVOIR ILLUMINATED.—SEE PAGE 342.

128 THE ILLUSTRATED LONDON NEWS JAN. 29, 1870

THE SEWERS OF PARIS.

THE BOAT.

THE WAGGON.

6.7 Visitors in the Paris sewers as depicted in the *Illustrated London News*, 29 January 1870, p. 128, showing women riding in boats (upper engraving) and men and women in carts (lower engraving).

appropriately heroic, while the illustration (**6.6**) shows a range of male figures, the bulk of whom are the top-hatted dignitaries led by Bazalgette, but with a group of three hatless figures on the right, possibly workers, and a policemen in the centre. This is an orderly group of men in a sublime but equally ordered space, visually stressed by the uniform lighting on the repeating brick arches which recede almost to infinity in the background. If some perceived this space as monstrous, it is here reassuringly pictured as the domain of masculine heroism, embodied in the figure of the engineer, who rationalises its spaces to his dignified visitors.

When the new Paris sewers were opened for public tours during the Exposition of 1867, both men and women visited these subterranean spaces, shown clearly in depictions in the *Illustrated London News* in 1870 (**6.7**). Visitors travelled through the sewers in deluxe versions of the vehicles used to cleanse them. In the top image women ride in a *bateau-vanne* with the men following them alongside on the wide walkways, while in the image below men and women ride together in the *wagon-vanne*, specially adapted for visitors. Unlike at Crossness, visitors rode through the Parisian sewage itself which, despite being absent of human excrement until the end of the 19th century, was nevertheless significant in its dirty symbolism. If visitors were impressed by the tidiness and order of the new system, by the dazzling illumination, or by the presence of lovely women dressed in their finery, they also read this experience in the light of one particular literary precedent: the climactic moment in Victor Hugo's *Les Misérables*, published in 1862, where the hero Jean Valjean hides in the Paris sewers after the abortive revolt of 1832. However, Hugo's sewers were very different from their 'clean … and correct' modern counterparts: they are 'excremental crypts' with 'poisonous ooze on the walls, drops falling from the roof, darkness'.[43] Yet, these ancient sewers also save Hugo's hero from the chaos of the above-ground world, protecting him in their womb-like spaces.

Returning to the ceremony at Crossness, the women who attended were elevated and segregated from the men (**6.5**), confined to the 'dainty palace of machinery' above-ground, and excluded from the subterranean reservoir, the space that came closest not only to the filthy sewage but also to older visions of a monstrous underworld. The transformation of the city's ancient sewer system into a rationalised space was enacted through the lens of a masculine audience, whether the engineer and his workers, the intrepid visitors seen in the engraving or the assumed middle-class male readers of the *Illustrated London News*. By contrast, in Paris, men and women alike could appropriate the experience of going underground. If the respectable women might have experienced the frisson of associations with 'ladies of the night' they might also have relived the romantic drama of Hugo's story, while the men would have appreciated the orderliness of the space but also perhaps, in their chivalric behaviour, an element of Valjean's heroism in the face of imagined dangers. The sewer tour might have domesticated these dangers but it also allowed an imaginative reconstitution of them from the point of view of both sexes within a reassuring context.

The question remains as to what extent either the perceived magical or monstrous nature of the spaces at Crossness and Abbey Mills was intentional on the part of Bazalgette or the architect, Charles Driver. In relation to the magical qualities of the interior of the Crossness engine-house, there is no documentation as to Bazalgette's opinions; neither is there any on his attitude towards the architectural flamboyance of Abbey Mills (see chapter 5). Nevertheless, Bazalgette would have approved all of the decorative details as well as the arrangements for the respective ceremonies; therefore he may have intended such imaginative associations, even if he does downplay this in his descriptions of the pumping stations. More importantly, in drawing attention to the synthesis of the artistic and the functional at Abbey Mills, press accounts confirm Driver's intentions as an architect, discussed in detail in chapter 5. Driver's own papers and architectural practice demonstrate his intention to imbue 'functional' materials, in particular cast iron, with aesthetic and moral principles, defined by Driver as 'Art', 'Sublimity' and 'Beauty', and derived from Ruskin's architectural 'lamps' (see chapter 5). Driver's application of these principles to cast iron was intended to transform conventional perceptions of it as prosaic and functional into something far more elevated: something magical or sublime. Press accounts of the interior of the Crossness and Abbey Mills engine-houses refer directly to the way in which the treatment of the ironwork contributed to the magical quality of their spaces: in relation to Crossness, some accounts describe the 'Byzantine arches' of the interior ironwork as singularly 'handsome'.[44] For the *Daily Telegraph* and the *Daily News*, even the brilliant colouring of the ironwork, stated to be in the style of Owen Jones, contributed to the magical quality of the interior space.[45] Likewise, at Abbey Mills, *The Times* compares the iron lantern to an ecclesiastical cupola,[46] while the *East London Observer* calls it a 'light and graceful dome'.[47] The *South London Press*, despite objecting to the costly decoration of this 'palace of filth', compares the plan of the building to a Greek orthodox church and describes a distinct 'tinge of Byzantine glamour' in the interior.[48] If the *East London Observer* notes this grumbling as to the expense, it counteracts it by highlighting the necessity of such ornamentation to make visible and celebrate the vast, but invisible, system of sewers to which Abbey Mills connected.[49] For most observers, the cast-iron architectural details of both the Crossness and Abbey Mills engine-houses are important and enriching elements in the magical atmosphere that pervaded the ceremonies, particularly at Crossness, and their observations confirm the importance of the architect in providing a dramatic 'backdrop' for these events.

In relation to the subterranean sewage reservoir at Crossness, Bazalgette was directly involved in both its design and decoration, being given responsibility for its illumination on the day of the ceremony.[50] Given his consistent rationalising of his project, Bazalgette's choice of decoration for the reservoir is surprising. For it was the lighting in particular that provoked the most comment by the press: if the thousands of lamps brought out the 'nicety [of] the brickwork'[51] they also created an atmospheric sense of obscurity – visitors appeared 'dim and shadowy;'[52] there was a magical 'play of light' in the arches;[53] and the 'Cimmerian gloom' appealed to those with a 'poetic turn of mind'.[54]

Furthermore, according to the writer for the *Daily Telegraph*, it was precisely the lighting that generated monstrous oppositions to the magical. Bazalgette possibly intended the space to appear magical or pleasing; almost certainly, he did not envisage it being cast as monstrous. Perhaps the lack of such associations in press accounts of the ceremony at Abbey Mills three years later points to a deliberate attempt by Bazalgette to more effectively control the responses to the building, by guiding visitors with a fully rationalised description of its spaces. Nevertheless, the sense of the monstrous in the space of the Crossness reservoir remains a significant aspect of some of the press' response and it requires further exploration: the remaining part of this chapter will asses how this monstrous perception fits into the wider context of ideas about sewers in the 1860s.

Sewers and the Monstrous

In April 1865, anticipating the ceremony at Crossness, the leading daily newspapers – *The Times*, *Standard* and *Daily News* – published related articles on the 4th and 5th: the former detailing Bazalgette's main drainage system and the old sewers and cesspools it superseded; the latter concentrating on the ceremony itself. Many of the articles on 4 April directly compare the new system with the old: if Bazalgette's sewers are stupendous, marvellous and mythic in their importance, they replaced something starkly different: an ancient, dilapidated system of sewers overloaded with 'pent-up refuse'[55] and cesspools filled with 'monstrous impurities' and symbolised by the 'disgusting occupation of the nightmen' (workers who emptied the cesspools at night).[56] According to these accounts, the 'monstrous evil' that was the old system is remedied by Bazalgette's new sewers and pumping stations.[57] Here it is old sewers and cesspools that are configured as monstrous, with the new system replacing such associations with something far more elevated.

However, in the space of the reservoir at Crossness, some members of the press, especially the writer for the *Daily Telegraph*, experienced a temporary reversal of these conventional associations, when the magical 'new' took on a monstrous aspect more characteristic of the 'old'. Furthermore, it was paradoxically the magical quality of the space, provided by Bazalgette himself, that generated – in this writer – monstrous counter associations: here it seemed, for a moment at least, old and new were disturbingly confused.

Such configurations of old and new, often in similar surprising reversals, are common in other accounts of London's sewers published during the planning and construction of the main drainage system in the 1850s and 1860s. Throughout this period, engineers and other 'experts' challenged the basic principles of Bazalgette's scheme (see chapter 2 for more detail). Pamphlets by George Booth, George Rochfort Clarke and John Wiggins represent a sample of these voices of resistance, which continued even after Bazalgette began to construct his intercepting sewers in 1859.[58] All of these pamphlets reassert the importance of recycling London's sewage – an imperative originally propagated by Chadwick and engineers associated with him but subsequently underplayed by Bazalgette (see chapter 2).[59] In a dramatic reversal of press perceptions of the main drainage system, these accounts configure the old sewers and cesspools as 'safer and sweeter' than Bazalgette's new 'monstrous' sewers.[60] Clarke proposes that the old system of sewers be retained, castigating the Metropolitan Board of Works for simultaneously throwing away a valuable resource and creating a 'monster grievance' further downriver. Bazalgette's intercepting sewers are variously cast as an unmanageable underground 'labyrinth', an 'entangled net', or the 'deformed and barren offspring of water-closets'.[61] With equally invective language Wiggins and Booth argue along similar lines: Booth sees Bazalgette's 'monster sewers' as contributing to, rather than eradicating, the spread of disease; because of their enormous size the 'pestilential vapours' inside cannot be controlled.[62] Wiggins echoes these concerns, calling Bazalgette's plan 'alarming' and 'deranged'.[63] Booth's main fear is the concentration of dangerous gases in Bazalgette's enormous sewers and 'vast' reservoirs, like that visited at Crossness in 1865.[64] He goes on to propose that:

> Instead of building the Herculean works and stupendous erections proposed, it would appear that the sewers of London may, like the city itself, be so constructed in integral parts, small in their separate divisions ... yet effective as a whole.[65]

Reversing the positive meaning of 'stupendous' in the press accounts, Booth reasserts the 'old' vision of London and its sewers that had been so effectively condemned by those who first envisaged a comprehensive sewer system for London in the late 1840s (see chapter 1).

If this inflammatory language might be put down to rivalry within the narrow confines of London's engineering community, other responses suggest a wider disruption of the old and the new in regard to the main drainage system. The celebrated journalist Henry Mayhew's (1812-87) encyclopaedic account of London's underclass, *London Labour and the London Poor* (1862), contains a description of the city's sewers, both past and present that is remarkable for its rich attention to detail:[66] it includes descriptions of all the different types of sewers in the city, their sizes, lengths and condition, as well as a long account of the exact constituents of the sewage itself. In a section dealing with the history of the administration of London's sanitation, Mayhew directs his invective language against the old 'wretched' system, describing, in straightforward condemnatory language, the terrible condition of the old sewers of the city and the pressing need for an entirely new system.

However, when Mayhew gives a romanticised account of an evening spent with a gang of London nightmen, he 'registers his resistance to a reforming process' that sought to obliterate the old system and its attendant workers.[67] Equally ambivalent are Mayhew's descriptions of other sewer 'workers', such as the sewer hunters (known as toshers) and the mud-larks, both of whom exploited London's old drainage system by scavenging for discarded valuables in its spaces. According to David Pike, in his descriptions of these marginal social characters Mayhew 'implies the eventual disappearance of these 'outcast' occupations' and displays a 'Romantic nostalgia' for their activities.[68] Significantly, Mayhew's accounts had a long publication history: they were first collected in the late-1840s and published in the *Morning Chronicle*, then republished as three volumes from 1850 to 1852 and expanded into four in 1862. Throughout this period, London's main drainage system was first planned and then implemented; the definitive 1862 volumes of *London Labour and the London Poor* were published at the peak of activity in its construction. Yet Mayhew fails to mention Bazalgette's scheme and its huge impact on the city's sanitary infrastructure. Rather, his vivid cast of sewer dwellers and workers – by then very much consigned to the past – remain in his volumes as real and tangible presences in the city's substructure, emphasised by the wood-engraved illustrations of these workers included in the volumes and supposedly based on photographs.[69] In short, Mayhew oscillates between seeing the old London sewers as monstrous but also as a romantic remnant of a fast-vanishing pre-industrial society.

Ambivalence also characterises what is perhaps the most obsessive engagement with London's sewers in the 1860s: John Hollingshead's *Underground London*, published in 1862.[70] Following Mayhew's journalistic pretext, Hollingshead's series of essays derive from the author's self-confessed 'appetite for the wonderful in connection with sewers'.[71] Of the seventeen chapters that comprised *Underground London*, nine directly concern sewers, whether describing the old system, journeys within the sewers themselves, or Bazalgette's main drainage system, then under construction. Hollingshead's collection of essays is remarkable for the sheer variety of viewpoints represented. Indeed, he sums up these multiple conceptions of sewers in his introductory chapter:

> There are more ways than one of looking at sewers, especially old London sewers. There is a highly romantic point of view from which they are regarded as accessible, pleasant, and convivial hiding-places for criminals flying from justice, but black and dangerous labyrinths for the innocent stranger ... [and] there is the scientific or half-scientific way, which is not always wanting in the imaginative element.[72]

The subsequent chapters of *Underground London* present London's sewers, old and new, in simultaneously rational, magical and monstrous terms – terms that play off the supposed oppositions of old and new in more conventional discourse. In chapter 4, Hollingshead presents the old sewers as a rational and therefore legible system, the author providing the reader with a 'panorama' of their spaces; while chapters 5 and 6, describing his journeys within these sewers, configure them paradoxically as repositories of the monstrous (dead bodies and ferocious rats) and the domestic (genteel sewer workers). If, in chapter 5, Hollingshead views Bazalgette's main drainage system as a 'great accomplished fact' and a successful example of the 'struggle of art against nature', he also states that its vast intercepting sewers, which 'dwell in perpetual darkness', might be seen by some as 'volcanoes of filth; gorged veins of putridity; ready to explode at any moment in a whirlwind of foul gas, and poison all those whom they fail to smother'.[73]

Conclusion: Spatial Conflation

The pervasiveness of configurations of old and new in relation to London's sewers in the 1860s – in both press accounts of the ceremonies at Crossness and Abbey Mills and in wider discourse – is not surprising given the very tangible transformation of the city's sanitary infrastructure that occurred in this period. The modernising of London in the 1860s, of which the construction of the main drainage system was an important part, was perceived by contemporaneous commentators as both a rationalising process and also as a fertile stimulus for the imagination. If the pumping stations were, according to Bazalgette, rational 'icons of the new' they were equally, for some, possible sites for older 'dreams, memories and fantasies'.[74] The press accounts examined in this chapter demonstrate the extent to which the old is perpetually engaged with the new: in this case in a shifting dialogue between the rational, the magical and the monstrous.

The main drainage pumping stations were intended by their creators (engineer and architect) to be visible symbols of a vast new underground system of sewers – a system built at great cost in order to transform the old and 'monstrous' sanitation of the city. In line with this symbolic status, the pumping stations were embellished with lavish architectural decoration, elevating their value above mere utility and imbuing their spaces with a sense of sublime nobility. This was, in effect, an entirely new vision of sewers, where the magical was overlaid onto the rational in order to make the prosaic pleasing. However, in the particular space of the subterranean reservoir, where even if the sewage was not visible it was however present in the imagination, older associations emerged, or perhaps resurfaced, with the writers fully alert to the contradictory aspects of this experience and the pull of both fascination and fear. It was the unique character of this space that prompted a conflation of the rational, magical and monstrous – a point emphasised by the writers who knew this was an experience never to be repeated. Indeed, for the writer for the *Daily Telegraph*, the new vision paradoxically stimulated these older associations: in short, old and new were inseparably linked. As if to exemplify the unique character of this experience, three years later, in the ceremony at Abbey Mills, visitors also wondered at the lavish decoration and vast machinery but, guided by Bazalgette's fully rationalised account, did not refer to any monstrous associations. Indeed, even when presented with the opportunity of inspecting the sewage pumps below ground, most of the visitors declined;[75] even the *Daily Telegraph*, whose correspondent had, three years earlier, been so rampant in his imaginative prose, gave little attention to these 'noisome chambers far below' the building's lavish interior.[76]

Were these visitors in 1868 now completely won over to the new vision of sewers? Or did the engineer more effectively control their responses? Certainly, the *Daily Telegraph*'s response to the Crossness reservoir, three years earlier, with its conflation of the new (the rational and the magical) and the old (the monstrous) demonstrates that, at particular times and in particular places (and perhaps for particular people), rationality and imagination and old and new might be configured in unexpected hybrid forms. Such a conclusion brings this book full circle: from the uneasy conjunction of old and new in the first stages of planning the main drainage system to just such a conjunction in its reception some 20 years later.

Postscript

Today, public access to the London and Paris sewers remains remarkably similar to that seen in the late-1860s. In London, Thames Water's annual 'Open Sewers Week' in May uses the Abbey Mills pumping station as a base for a tour of the building, a lecture on the history and future of London's sanitation, and an underground walk through a small section of the northern outfall sewer. Meanwhile in Paris, the Musée des Egouts near the Pont de l'Alma has made the Paris sewer tour a regular feature of any tourist's sightseeing itinerary. Yet, the spaces of the sewers themselves could not be more different. In Paris, visitors wander unaccompanied along wide walkways through cavernous spaces and are reminded of Victor Hugo's influence by means of rather crude illustrations depicting scenes from *Les Misérables*. London's sewers still appeal to the more intrepid, but not necessarily male visitor; during Open Sewers Week, both men and women are lowered by a rope into the northern outfall sewer

P.1 Walking through the northern outfall sewer during Thames Water's 'Open Sewers Week', May 2007.

in protective clothing, lead-weighted boots and an oxygen tank in case of poisonous gas. The interiors of the London sewers are no less hostile; the visit entails wading through knee-deep sewage along a smooth brick-lined tunnel in almost complete darkness (**P.1**).

If London's sewers are now almost completely hidden from public view, they are occasionally made visible to provoke a sense of wonder like that experienced by the first visitors to Crossness and Abbey Mills in 1865 and 1868. The recent BBC film 'The Sewer King' was part of its series *The Seven Wonders of the Industrial World* (2003). As one of these 'wonders', Bazalgette's main drainage system was visualised using computer-generated images which recreated the process of construction from contemporaneous images, many of which have been reproduced in this book. The film was clearly intended to celebrate and memorialise Bazalgette and his most important achievement, confirmed by the fact that it was produced by one of his descendents, Edward Bazalgette. Such a sense of sewers as wondrous goes against the grain of popular perceptions of them as unpleasant spaces that contain the most horrible of wastes; it is the latter perception that has dominated imaginative uses of sewer spaces in film and literature. Carol Reed's 1949 film *The Third Man* explores human depths – unconscious motives, hidden political and personal treachery, and death – which are symbolised by, and return through, the ultra-rationalised spaces of the Vienna sewers just after the Second World War. In a different vein, the smell of London's sewers summons up childhood memories for the female protagonist of Margaret Drabble's 1980 novel *The Middle Ground*: stooping to take a sniff at a grating ventilating one of Bazalgette's sewers, she cannot resist this 'powerful odour of London' that invites her to 'escape the prison of the present into the past, where dark spirits swam in the fast-moving flood'.[1]

Things that bubble up from the unconscious might be altogether more unpleasant and, in the world of film, sewers have provided popular locations for nightmarish monsters: from giant ants in the storm drains of Los Angeles in the cold-war thriller *Them!* (1954); mutant alligators in the sewers of Chicago in *Alligator* (1980); to more recent incarnations such as human-like cockroaches in Guillermo Del Toro's 1997 film *Mimic*. Throughout the post-War period the imaginative connotations of London's sewers have tended to be displaced by those of other cities, in particular New York; yet recently they have resurfaced in both literature and film. In the final moments of Peter Carey's 1997 novel *Jack Maggs* the eponymous hero witnesses the construction of the city's Victorian sewers. Here, the 'vertiginous unease' induced by the sight of a deep trench being dug in the street mirrored the general anxiety Jack Magg's felt about his own life and summoned up an apocalyptic vision of his own demise.[2] Likewise Clare Clark's 2005 novel *The Great Stink* sets most of its narrative in the London sewers, exploiting their dark associations to mirror the repressed yearnings of her central character, which are played out in the hidden spaces of the sewers before dramatically entering the life of the world above. More visceral still is the brutally feral monster inhabiting a self-made netherworld in Christopher Smith's 2004 film *Creep*, who, at night, returns from the sewers through the tunnels of the Underground to enact vicious killings. Although crass and exploitative, the horrors in *Creep* seem to prefigure the much more tangible sense of unease now associated with the city's substructure since 7 July 2005. Engineers of the past and present might build sewers as rational spaces that bring wastes to order, but it seems they will always be open to other subversive interpretations and uses; clearly we are still fearful of what terrors might return to confront us from the darkness of the world below.

Notes

Abbreviations

LMA London Metropolitan Archives
MBW Metropolitan Board of Works
TWA Thames Water Archive, Abbey Mills Pumping Station, London

Introduction

1. *Observer*, 14 April 1861, p. 5, 'The Metropolitan Great Drainage Works'.
2. John Hollingshead's series of essays appeared in *All the Year Round* as follows: 26 January 1861, no. 92, pp. 453-6, 'Underground London. Chapter III'; 20 July 1861, no. 117, pp. 390-4, 'Underground London. Chapter I'; 27 July 1861, no. 118, pp. 413-17, 'Underground London. Chapter II'; 10 August 1861, no. 120, pp. 470-3, 'Underground London. Chapter IV'; and 17 August 1861, no. 121, pp. 486-9, 'Underground London. Chapter the Last'. The essays were collectively published in book form as *Underground London* (London, 1862).
3. Hollingshead, 'Underground London. Chapter the Last', p. 486.
4. Hugo, p. 1064.
5. On the history of domestic sanitation see Horan; Pudney; and Wright.
6. On the history of the Thames Embankment and its connection with the main drainage system see Bazalgette (1868), pp. 3-4; Halliday, pp. 144-63; Owen, pp. 74-101; and Dale Porter. Porter's book-length study remains the definitive account.
7. For lists of the original staff at the Deptford, Crossness and Abbey Mills pumping stations see LMA, MBW, Minutes of Proceedings, 2 March 1866, p. 281, s. 8; 2 March 1866, pp. 279-80, s. 8 and 18 May 1866, p. 667, s. 6; and 18 December 1868, pp. 1358-9, s. 8 respectively. On the community of workers at Crossness see *The Crossness Engines* (London, 2002), pp. 23-5.
8. On the Paris sewer workers see Reid, pp. 87-178.
9. Jephson, pp. 158-9.
10. See Clifton on the development of London's government in the 19th century; Hamlin (1998) for analysis of the debates amongst sanitary reformers in the 1850s; Luckin on the river Thames in the 19th century; and Wohl on the emergence of ideas of public health in the same period.
11. These accounts include Clayton; S. Smith; F.L. Stevens; and Trench and Hillman.
12. Pike (2005), pp. 190-269.
13. Nead, p. 24.
14. Ibid., pp. 24-9.
15. Allen, p. 2.
16. Certeau, pp. 91-4 and Lefebvre, pp. 38-9. While Certeau concentrates on the differences between 'planned' and 'lived' spaces, Lefebvre develops this disjunction into a triad of 'conceived', 'social' and 'representational' spaces.
17. Ben Johnson, 'On the Famous Voyage' (*c.*1610) in C.H. Herford et al.
18. The 1801 census stated the population of London as 959,000; the 1851 census stated it as 2,362,000.
19. For more detailed accounts of London's natural drainage see Barton; and Halliday, pp. 17-34.

20. For a description of the work of the London nightmen see Mayhew, vol. 2, pp. 451-2.
21. Wright, pp. 107-8. Joseph Bramah's valve closet, patented in 1778, became the accepted model for water closet production in the 19th century. By 1797 he had already made approximately 6,000 closets.
22. Parliamentary Acts, 4 & 5 Will. IV, c. 76.
23. Owen, p. 33.
24. Halliday, pp. 102-7.
25. Trench and Hillman, p. 76.
26. On these debates see Halliday, pp. 108-23.
27. Halliday, pp. 105-6.
28. Ibid., p. 107.

Chapter 1

1. Harley (1988), p. 278.
2. On the Ordnance Survey of London see Darlington; Pinkus, pp. 124-34; and Nead, pp. 18-22. For more information on the subterranean survey see Darlington.
3. Darlington and Howgego's catalogue of London maps, by no means exhaustive, lists 422 separate maps produced from 1553 to 1850, 222 of which cover the period 1800-50.
4. Pinkus, p. 108.
5. For the history of the Ordnance Survey see Skelton, pp. 415-16; Close; and Owen and Pilbeam.
6. *Parliamentary Papers*, 1847-48, vol. 32, 'First Report of the Commissioners Appointed to Inquire Whether Any and What Special Means may be Requisite for the Improvement of the Health of the Metropolis' (hereafter 'First Report of the Commissioners'), p. iii.
7. Halliday, pp. 32-4. The seven amalgamated Commissions of Sewers were: the Crown-appointed Commissions of Westminster, Surrey and Kent, Holborn and Finsbury, Poplar, St Katherine's, Tower Hamlets, and Greenwich. The eighth – the City Commission appointed by the City Corporation – remained independently governed.
8. 'First Report of the Commissioners', pp. 3-47.
9. Ibid., pp. 48-52.
10. LMA, MCS/481: Metropolitan Commissions of Sewers, Minutes of Proceedings, 6 December 1847, p. 2.
11. MCS/481, 13 January 1848, pp. 7-8.
12. For the origins of the Corps of Royal Sappers and Miners see Connolly, vol. 2.
13. *Examiner*, 14 July 1849.
14. *Illustrated London News*, 24 June 1848, p. 414, 'Ordnance Survey of London and the Environs'.
15. *Builder*, 17 June 1848, pp. 291-2 and *The Times*, 4 November 1848, p. 5.
16. Nead, p. 17.
17. Owen, p. 33. Owen uses the parish of St Pancras as an illustration: in 1855, 19 separate Boards administered the parish, 16 for lighting and paving (three for lighting only). The 19 Boards comprised some 427 commissioners, 255 of them self-elected.
18. *Parliamentary Debates*, 3rd Series, vol. 97, 24 March 1848, p. 1016.
19. Parliamentary Acts, 6 & 7 Will. IV, c. 96.
20. MCS/476/16.
21. Parliamentary Acts, 11 & 12 Vict., c. 112.
22. Pike (2005), pp. 8-12. On the London panorama at the Colosseum see Oettermann, pp. 132-40.
23. Oettermann, pp. 135-7.
24. MCS/481, 13 January 1848, pp. 7-8.
25. Ibid., p. 8.
26. Pike (2005), pp. 9-10.
27. MCS/481, 13 January 1848, p. 7.
28. Ibid., pp. 1-2.
29. MCS/487, 21 August 1855, p. 374.

30. Darlington, p. 62.
31. LMA, MCS/498/1-1384. The 1,384 notebooks held at LMA are an invaluable record of the surveying techniques of the levellers. Similar notebooks produced by the Ordnance Survey levellers in 1848 and 1849 were destroyed when the Ordnance Survey's offices in Southampton were bombed during the Second World War.
32. LMA, MCS/498/45: Westminster Levels, Book 45, p. 11.
33. See LMA, MCS/476/22, MCS/476/27, MCS/476/28, MCS/477/31, MCS/477/32, and MCS/481 for additional reports submitted for the week ending 8 August 1848 and from 8 August to 2 September 1848.
34. LMA, MCS/477/32, p. 11. On 7 December 1848, Smith's observation on the decrepit state of the sewer under Hinde Street included a cross-sectional drawing of the sewer.
35. MCS/477/31, p. 1.
36. Ibid.
37. LMA, MCS/476/27: Subterranean Survey: Report to Survey Committee from the 2nd to 16th September, 1848 (Henry Austin).
38. MCS/477/32, 7 December 1848, p. 11.
39. Ibid., 28 October 1848, p. 9.
40. Ibid., 7 December 1848, p. 11.
41. Ibid., 28 October 1848, p. 9.
42. LMA, MCS/476/27.
43. LMA, MCS/189: Minute and Report Book, Metropolitan Sewers, Ordnance Survey Committee. The Ordnance Survey Committee, formed on 16 January 1849, was the new title given to what was the Sub-Committee of Works for the Ordnance Survey, originally set by Chadwick in January 1848.
44. MCS/189, 8 June 1849.
45. Darlington and Howgego, p. 28.
46. While there are no indicators in the minutes of the meetings of the Metropolitan Commission of Sewers as to why these maps were abandoned, Chadwick's departure from the Commission in September 1849 probably precipitated a change in policy on the part of the commissioners. In 1853, Smith did produce an experimental sheet on the ten-feet to one-mile scale depicting house drainage in Toothill Street, Westminster. However, it remained an isolated attempt to map the details of house drainage. See LMA, MCS/P/25/ 1-15 (1) for this sheet.
47. Darlington, p. 63. As Darlington has noted, the production of these sheets formed a precedent in that it influenced the decision by the Ordnance Survey to produce ten-feet to one-mile plans of other large towns.
48. LMA, MCS/P/25/1-15 (10). This sheet was more highly finished than **1.5**.
49. MCS/189, 1 June 1849.
50. Sewer information was eventually hand-drawn onto the five-feet to one-mile sheets of the Ordnance Survey (as seen in **1.2**). This process occurred at a later date to the subterranean survey – probably sometime from the mid-1850s to the re-mapping of London by the Ordnance Survey in the late 1880s. Over 300 of these maps are currently held in the Thames Water archive at the Abbey Mills pumping station in London.

Chapter 2

1. Schivelbusch, p. 263.
2. For accounts of this process see Finer, pp. 297-474; Halliday, pp. 48-76; Hamlin (1992); and Owen, pp. 31-73.
3. LMA, MCS/481, 6 December 1847, p. 2.
4. Flinn, pp. 98, 109-10, 164-6, 282, 375-8, 423-4.
5. Ibid., pp. 339-410.
6. *Parliamentary Papers*, 1844, vol. 17, 'First Report of the Commissioners for Inquiry into the State of Large Towns and Populous Districts' (hereafter 'Health of Towns Report'), pp. x, xi.
7. Hamlin (1992), p. 684.

8. On the interest in sewage utilisation in the 19th century see Goddard; Halliday, pp. 108-23; Hamlin (1989); Sheail; and Smith and Young.
9. See *Parliamentary Papers*, 1852-53, vols 19 and 26; 1857-58, vol. 32; 1865, vols 8 and 27; and 1884, vol. 41 for a representative sample of these reports.
10. Liebig, p. 202.
11. Quoted in Hamlin (1989), p. 97.
12. Hugo, p. 1061. On Hugo and sewage recycling in France see Reid, pp. 54-7; and Pike (2005), pp. 200-3.
13. Quoted in Finer, p. 222.
14. Hamlin (1989), p. 95.
15. Ibid, pp. 102-3.
16. Ibid, p. 105.
17. *Parliamentary Papers*, 1852, vol. 19, 'Minutes of Information Collected with Reference to Works for the Removal of Soil Water or Drainage of Dwelling Houses and for the Sewerage and Cleansing of the Sites of Towns' (hereafter 'Minutes on House Drainage').
18. Ibid., p. 144.
19. Ibid., p. 98.
20. Ibid.
21. Hamlin (1992), p. 683; Lewis, pp. 52-4; and Osborne, p. 114.
22. Osborne, pp. 113-14.
23. 'Minutes on House Drainage', p. 107.
24. Ibid., pp. 14, 62.
25. Ibid., p. 111.
26. Ibid., pp. 111, 125-37.
27. See Hamlin (1992), p. 698. 100,000 copies of the 'Minutes on House Drainage' report were distributed to the local Boards of Health in British towns proposing the adoption of the principles laid down.
28. 'First Report of the Commissioners', p. 118.
29. Ibid.
30. Ibid., p. 120.
31. See *Parliamentary Papers*, 1857, vol. 20, 'Report on the Means of Deodorizing and Utilising the Sewage of Towns', p. 3. Austin states here that 'the great cycle of life, decay, and reproduction must be completed, and so long as the elements of reproduction are not employed for good, they will work for evil'.
32. 'First Report of the Commissioners', p. 124.
33. Ibid., p. 122.
34. Ibid.
35. Ibid., p. 127.
36. Ibid., p. 131.
37. LMA, MCS/482: Metropolitan Commissions of Sewers, Minutes of Proceedings, 17 October 1849, p. 3. Austin went on to become secretary to the General Board of Health.
38. Ibid., 20 August 1849, p. 219.
39. Ibid., 16 August 1849, p. 213.
40. See Halliday, pp. 1-16, and Smith (1991), p. 3 for biographical information on Bazalgette.
41. LMA, MCS/478/3: 'Report on the Plans for the Drainage of London' (15 March 1850), pp. 2-3. The seven categories were: portable cesspool systems; schemes that proposed the Thames as the principal outlet for sewage; single tunnel sewer schemes; intercepting systems of more than one sewer; schemes with cesspool outlets on the banks of the Thames; converging systems; and systems that improved the existing drainage network.
42. Ibid., p. 2. The 1852 Metropolis Water Act, which required the London water companies to draw their water from the Thames above the tidal reach at Teddington Lock, recognised the dangerous extent to which sewage had polluted the river in the built-up areas of the city.
43. MCS/478/3, p. 4.
44. Ibid., p. 3.
45. Ibid., p. 5.

46. LMA, MCS/483: Metropolitan Commissions of Sewers, Minutes of Proceedings, 25 January 1850, p. 91.
47. See 'Mr. Frank Forster', *Civil Engineer and Architect's Journal,* 15 (1852), p. 160; and *Minutes of Proceedings of the Institution of Civil Engineers* 12 (1852-3), pp. 157-9 for detailed biographies of Forster.
48. LMA, MCS/478/4:'The Engineer's Report on the Surrey and Kent Drainage' (1 August 1850), p. 1.
49. Ibid.
50. Ibid., p. 3.
51. Ibid., p. 6.
52. Pike (2002), p. 104.
53. Ibid. For further detail on Napoleon III's map see also Jordan, pp. 170-6; and Pinkney, pp. 25-31.
54. Pike (2002), p. 105.
55. *Civil Engineer and Architect's Journal*, 15 (1852), p. 160. Forster subsequently died on 13 April 1852 apparently 'in the act of writing a letter when he was struck with apoplexy, and almost immediately expired'.
56. LMA, MCS/486: Metropolitan Commissions of Sewers, Minutes of Proceedings, 26 November 1852, p. 90.
57. LMA, MCS/479/3: 'Report upon the Sewage Interception and Main Drainage of the Districts North of the Thames' (21 January 1854), p. 17.
58. Ibid., p. 5.
59. LMA, MCS/479/13: 'Report by J. W. Bazalgette on the Sewage Interception and Main Drainage of the Districts North of the Thames' (22 May 1856), appendix B.
60. Ibid., p. 10.
61. Ibid., p. 8.
62. See the extensive collection of local sewer proposals held at the Thames Water archive, Abbey Mills pumping station. From the 1860s onwards all proposals for local sewers were checked by the MBW, most by Bazalgette himself.
63. MCS/479/13, p. 4.
64. Ibid., p. 5.
65. MCS/481, 13 January 1848, pp. 7-8.
66. LMA, MCS/479/2:'Report on the High Level Line for the Interception of the Drainage North of the Thames and on the Intercepting Lines South of the River' (6 October 1853), p. 19.
67. MCS/479/13, pp. 11-12.
68. MCS/479/3, p. 19.
69. Ibid., pp. 5-6.
70. Ibid., p. 31.
71. LMA, MCS/479/4: 'Report by Robert Stephenson on the Sewage Interception and Main Drainage of the Districts North of the Thames' (15 May 1854), p. 4.
72. LMA, MCS/479/5:'Report to the Metropolitan Commission of Sewers on the Northern Drainage by John Roe' (23 October 1854), pp. 3-12.
73. Ibid., pp. 17-46.
74. LMA, MCS/479/6: 'Further Report to the Metropolitan Commission of Sewers in Respect of that Portion of the Main Intercepting Drainage Called the Northern Drainage' (16 January 1855), p. 3.
75. LMA, MCS/479/7: 'Data, Employed in Determining the Sizes and Estimating the Cost of the Works Designed for the Main Drainage of the Metropolis' (22 January 1855), p. 3.

Chapter 3

1. See Ferguson, whose focus on the cognitive processes of the engineer follows a similar emphasis seen in Baynes and Pugh; Booker; and Hindle.
2. Brown (1999); Brown (2000); Lefèvre; Lubar; McGee; and E.W. Stevens. A range of new analytic approaches are offered by these sources: in both of his articles Brown focuses attention on the development and uses of mechanical drawing in the 19th century in

order to reveal the 'cultural beliefs and political ends of engineers' (Brown (2000), p. 196); Lefèvre and others focus on the significance of drawings in the practice of engineers and architects in the early modern period, emphasising the variety of their communicative purposes and related audiences; Lubar's analysis centres on the role played by technical representations in the power relationships bound up with the implementation of any technology; McGee sets out 'formal criteria for distinguishing among the technical activities of the past' (p. 210), focusing on British naval architecture in the early modern period; while Stevens's book charts the rise of a new technical literacy in the United States in the first half of the 19th century, bringing out the role of technical representations as both tools of design innovation and also important players in the establishment of a culture of technical education.

3. The most comprehensive accounts of the construction of the main drainage system are Halliday, pp. 86-90; and Owen, pp. 56-61. Both Halliday and Owen's accounts refer to the importance of the contract in the construction process, but both display a conventional understanding of the contract drawings and specifications as cognitive tools of the engineer. Consequently, neither explores the wider social context in which these representations were situated.
4. Metropolitan Local Management Act, 1855, 18 & 19 Vict., c.120, clause 135. This clause states that the MBW had the duty of constructing 'such sewers and works as they may think necessary for preventing all or any part of the sewage within the Metropolis from flowing or passing into the River Thames in or near the Metropolis'.
5. LMA, Minutes of Proceedings of the MBW (hereafter MBW Minutes), 3 July 1856, pp. 293-4, s. 3.
6. On the referees plan see Halliday, pp. 99-103, and Owen, pp. 51-5. The MBW Minutes from 1856 to 1858 also provide an exhaustive record of this dispute.
7. LMA, MBW/076: 'Main Drainage of the Metropolis. Report to the Metropolitan Board of Works upon the Main Drainage of the Metropolis' [J. W. Bazalgette with George Bidder and Thomas Hawksley] (6 April 1858).
8. On the 'Great Stink' see Clifton, pp. 23-4; Glick; Halliday, pp. 104-11; Horrocks; Owen, pp. 53-4; and Dale Porter, pp. 70-2.
9. Metropolitan Local Management Amendment Act, 2 August 1858, 22 & 23 Vict., c. 104, clauses 1, 4, 6, 10, 12, 23.
10. *Parliamentary Papers*, 1888, vol. 56, 'Interim Report of the Royal Commissioners Appointed to Inquire into Certain Matters Connected with the Working of the Metropolitan Board of Works, Minutes of Evidence taken on the Metropolitan Board of Works Enquiry Commission' (hereafter 'Interim Report of the Royal Commissioners').
11. Bazalgette awarded the Thames Embankment Contract No. 1 (1864) to George Furness after the contract had only three weeks earlier been awarded to the lowest tender Samuel Ridley. For additional information on the Thames Embankment contracts see Owen, pp. 84-6.
12. MBW Minutes, 4 August 1865, p. 984, s. 30; and 20 April 1866, pp. 489-91, s. 4.
13. 'Interim Report of the Royal Commissioners', p. 321, Bazalgette's evidence of 10 July 1888.
14. See LMA, MBW/2421/1-11, Main Drainage Contracts Nos 1-11.
15. MBW Minutes, 12 August 1859, p. 578, s. 9.
16. MBW Minutes, 10 August 1860, p. 594, s. 36.
17. On the evolution of engineering contracts see Halliday, pp. 84, 86; Middlemas, pp. 171-3; and Thompson, pp. 66-82.
18. See Cooney, p. 157; Halliday, p. 84; and Summerson (1973), pp. 12-13. Halliday cites the role of Thomas Cubitt (1788-1855) in providing the model for large-scale contracting firms. In 1851, according to Cooney, five large contractor firms, each employing more than 350 workers, dominated London's building industry. By 1873 there were six firms with more than 800 employees and 41 with more than 200.
19. See Owen, p. 56, and Halliday, p. 89.
20. MBW Minutes, 10 August 1860, p. 594, s. 36.

21. 'Interim Report of the Royal Commissioners', p. 341, Bazalgette's evidence of 13 July 1888.
22. MBW Minutes, 19 October 1860, pp. 709-10, s. 4.
23. Brassey was awarded the contract for the northern middle level sewer after the failure of Rowe; Webster, the southern high-level sewer; Dethick, the Ranelagh storm relief sewer; Moxon, the northern high-level sewer.
24. Wadsworth, p. 6.
25. MBW Minutes, 26 October 1860, pp. 729-30, s. 2. Furness's sureties were the eminent engineer Sir Joseph Paxton (1803-65) and Messrs Smith and Knight, contractors for a portion of the Metropolitan Underground Railway, built from 1859 to 1863.
26. Wadsworth, p. 5.
27. MBW Minutes, 23 December 1859, p. 849, s. 36.
28. On the origins of three-view orthographic projection see Lefèvre, pp. 209-44.
29. Ferguson, pp. 94-5.
30. For example see MBW Minutes, 9 November 1860, p. 786, s. 28. The Eastern Counties and London & Blackwall Railway Companies wrote a letter to the Secretary of State stating that they thought the purchase of their land for the northern outfall sewer contract was unnecessary.
31. See Clifton, p. 40. The MBW Minutes from 1859 to 1868 are filled with details of the multitudinous parties compensated and the disputes arising subsequently.
32. MBW Minutes, 21 December 1860, pp. 925-6, s. 4. See also MBW/2310: Annual Reports of the MBW, 10 August 1860, pp. 9-10. This report states that between 25 March 1859 and 25 March 1860, the MBW dealt with 40 claims for compensation to landowners for the northern outfall and high level sewer contracts alone, costing the Board £36,111 5s.
33. LMA, MBW/2310: Annual Reports of the MBW, 5 August 1859, p. 16. According to Bazalgette, 'the Board have successfully resisted all attempts to obstruct the progress of their works, several applications to the Court of Chancery for Injunctions to restrain their proceedings having been made and refused'.
34. Anon., *International Exhibition, 1862*, p. 38, 'Class 10: Civil Engineering, Architectural, and Building Contrivances; Sub-class B - Sanitary Improvements and Constructions', Entry 2369, 'Bazalgette, J.W. Spring Gardens. Drawings of, and Models Connected with the Metropolitan Main Drainage'.
35. *Engineer*, 27 September 1867, p. 272, 'Abbey Mills Pumping Station, No. V'. During August and September 1867, the journal published reproductions of nine drawings from the original contract.
36. See the *Illustrated London News*, 21 May 1864, p. 504, 'The Metropolitan Main-Drainage Works: Machines for Lifting the Sewage' (engraving) and 8 April 1865, p. 335, 'The Metropolitan Main Drainage Southern Outfall at Crossness' and 'Metropolitan Main Drainage: Plan of the Southern Outfall Works at Crossness' (engraving). These engravings are directly copied from drawings 1 and 9 respectively in the Southern outfall works contract (1862).
37. *Illustrated London News*, 28 May 1864, p. 513, 'London Main-Drainage Works: View and Section of the Outfall of The Northern Drainage at Barking Creek' (engraving).
38. LMA, MBW/2421/4: Contract No. 4, Iron Fencing, Northern Outfall Sewer, J. Horton, Iron Merchant (20 December 1859).
39. LMA, MBW/2421/20: Contract No. 20, Northern Outfall Reservoir, G. Furness, Contractor (27 February 1863).
40. LMA, MBW/2421/12: Contract No. 12, Northern Outfall Sewer, G. Furness, Contractor (6 December 1860), 'Specification of Works', pp. 3-71, and 'Provisions and Schedule of Prices', pp. 72-84.
41. D.W. Young was the quantity surveyor employed by Bazalgette. For one of Young's original lists of quantities see LMA, MBW/2429a: Bills of Quantities Prepared by D.W. Young, Southern Outfall Works, Low Level Bermondsey Branch, Low Level Main Line, 1862.
42. MBW/2421/12, 'Schedule of Prices', pp. 83-4.

43. MBW/2421/20, 'Specification of Works', p. 79, Clause 266.
44. Ibid., p. 70, Clause 229.
45. Ibid., p. 80, Clause 265.
46. Ibid., p. 74, Clauses 243 and 244.
47. Ibid., p. 73, Clause 241.
48. Ibid., p. 73, Clause 241, and p. 65, Clause 203.
49. See Helps; Middlemas; Wadsworth; and C. Walker. Middlemas provides biographies of Thomas Brassey, Sir John Aird, Lord Cowdray and Sir John Norton-Griffiths. Helps and Walker's accounts of the life of Thomas Brassey are exceptions to the general lack of biographical information on contractors. Wadsworth's account, as far as I know, is the only (brief) outline of the life of George Furness.
50. See Middlemas, p. 22.
51. MBW Minutes, 27 July 1860, p. 550, s. 22. For the reports see LMA, MBW/2321: Engineer's Monthly Reports, 1860-70.
52. LMA, MBW/2320: Engineer's Annual Reports 1861-1869 and MBW/2310: Annual Reports of the MBW, 1856-65. From 1856 to 1860 Bazalgette's engineering report formed part of the annual report of the Board. From 1861 to 1869 his report was printed separately.
53. For a complete list of the duties of the clerks of works see MBW Minutes, 23 October 1868, pp. 1183-85, s. 19; and MBW/2321, 3 November 1870, pp. 6-7.
54. MBW/2321, 1 February 1861, p. 6.
55. MBW/2320, 29 June 1861, p. 6 and MBW/2321, 4 April 1861. Furness was one of the first contractors to introduce mechanisation into the construction of civil engineering projects, using steam-cranes and concrete-mixing machines in the building of the northern outfall sewer.
56. MBW/2321, 6 June 1861, pp. 1-2 and 1 August 1861, pp. 1-2.
57. Ibid., 2 January 1862, p. 1.
58. Ibid., 6 February 1862, p. 1.
59. Ibid., 2 October 1862, pp. 1-2.
60. Ibid., 1 July 1863, pp. 1-2.
61. Ibid., 5 August 1863, pp. 1-2.
62. Ibid., 3 March 1864, pp. 1-2.
63. Ibid., 3 August 1864, p. 1.
64. LMA, 25.101 MET: MBW Reports, Main Drainage 1859-69: 'Report by the Engineer on the Increased Value of Materials and Labour used in the Main Drainage Works since the Year 1858' (3 June 1858), appendix, pp. 12-21. In this appendix, Furness, Brassey, Dethick and Moxon provided Bazalgette with excerpts from their accounts.
65. MBW/2320, 29 June 1861, pp. 3-4 and MBW/2310, 10 August 1860, p. 5.
66. 'Interim Report of the Royal Commissioners', p. 322, Bazalgette's evidence of 10 July 1888.
67. MBW Minutes, 10 February 1865, p. 198, s. 13.
68. Ibid., 17 February 1865, p. 232, s. 8.
69. One collection of seven photographs is held in Thames Water's archive at the Abbey Mills pumping station. Six of the photographs show various sites visited in 1861, 1862 and 1864. The seventh, by Negretti and Zamba, is part of a series commissioned by the MBW on the occasion of the opening of the Crossness pumping station on 4 April 1865. Another twelve photographs, part of the Otto Herschan Collection, are held in the online archive owned by Getty Images. Seven of these photographs show the construction of the northern outfall sewer, one of Crossness, and five of an accident that occurred in Shoreditch in 1862.
70. In July 1862 the Metropolitan Railway commissioned Henry Flather to photograph the construction of the company's underground railway from Paddington to Finsbury Circus. 65 photographs survive in the Henry Flather Collection, held in the Museum of London, while 28 photographs, alongside sixteen reproductions of drawings from the original contract, were published in Flather (1862).

Chapter 4

1. Taylor, pp. 434-4. Burke's influential essay *A Philosophical Enquiry into the Origins of Our Ideas of the Sublime and Beautiful*, first published in 1757, set out in a series of categories, the characteristics of the sublime and what might induce it (see Boulton, pp. 57-87). Subsequent literature on the sublime is enormous in its scope: for introductions to the sublime and aesthetics see Ashfield and de Bolla; and Hipple. For a contemporary assessment of the sublime see, for example, McMahon.
2. Boulton, p. 39.
3. Taylor, p. 434.
4. On the sublime in Turner's art see Wilton.
5. Bobrick, p. 75.
6. Klingender explores the relationship between representations of industry and the sublime (pp. 84-95). According to Klingender, depictions by artists such as George Robertson of the ironworks at Coalbrookedale, built in the 1750s in Shropshire, were important early examples of the evocation of industrial sites as sublime spectacles (pp. 86-9). More recently, Nye has outlined the importance of the sublime as a popular mode of experience in relation to the reception of new technology in America from the 1820s to the present day.
7. *Illustrated London News*, 8 July 1843, p. iv, 'Preface to Vol. II'. The early history of the *Illustrated London News* is outlined in De Maré, pp. 81-92; Fox, pp. 2-3; Houfe, pp. 67-74; and Jackson, pp. 291-313. Sinnema's book-length study deals more extensively with the history and agenda of the newspaper in the 1840s.
8. Most rivals to the *Illustrated London News* were less expensive but short-lived periodicals: the *Illustrated Times*, costing 2½d. (3d. after 1864), was founded in 1855 and ran until 1872; the *Penny Illustrated Paper* from 1861 to 1913; the *Illustrated Weekly News*, also priced at 1d., from 1861 to 1863 and from 1867 to 1869; and the *Illustrated News of the World*, costing sixpence, from 1858 to 1863. The publication of the first edition of the *Graphic* in 1869 marked the first serious competition to the circulation of the *Illustrated London News*. On the working methods of the wood-engravers of the periodical press see De Maré, pp. 13-92; Fox, pp. 1-13; Jackson, pp. 315-60; Klingender, pp. 65-70; and Sinnema, pp. 63-79.
9. *Illustrated London News*, 27 May 1843, p. 347, 'Our First Anniversary'.
10. Fox, p. 3.
11. *Illustrated London News*, 4 October 1845, p. 213, 'The Fleet-street Sewer'.
12. Ibid.
13. On the life and work of Frederick Napoleon Shepherd see Phillips. Thomas Hosmer Shepherd, father of Frederick and well known at the time for his topographical drawings of London, also sketched the scene in the same year. According to Phillips, in the 1840s, Frederick based many of his watercolours on his father's drawings (p. 107). Both images are held in the Guildhall Library Print Room, Corporation of London. It is not clear which of these images the *Illustrated London News* used as a basis for the engraving, but in the 1840s, the newspaper regularly included engravings copied from artists' work.
14. Taylor, pp. 444-5.
15. Williams, pp. 65-6, 88.
16. Potts, p. 36.
17. Phillips, p. 66.
18. Sinnema, p. 86.
19. Maidment, p. 145.
20. Bazalgette (1865), p. 314.
21. *Illustrated London News*, 19 February 1859, p. 173, 'The Metropolitan Main Drainage'.
22. *Illustrated London News*, 27 August 1859, p. 203, 'Main Drainage of the Metropolis'.
23. http://www.gettyimages.com/Search/Search.aspx?assettype=image&artist=Otto%20 Herschan#3. This online archive also includes eleven more photographs of the construction of Bazalgette's sewers held in the Otto Herschan Collection. It is unclear whether Herschan is the photographer or collector and whether or not they were commissioned images.

24. Engen, p. 244.
25. Martin and Francis, pp. 232-6. Although the 'wet-plate' process, developed in the 1850s, cut exposure time from minutes to a matter of 10 to 15 seconds, this still did not allow figures in motion to be depicted clearly.
26. *Illustrated London News*, 27 August 1859, p. 203.
27. Nead, p. 24.
28. Williams argues that middle-class attitudes towards the navvies centred on the notion of industry as 'heroic' and 'progressive' (pp. 52-63). Likewise, Klingender considers middle-class representations of navvies as expressing their 'heroism' (pp. 171-9). Berman's seminal study on modernity, *All that is Solid Melts into Air*, also refers to the importance of the workers in the eyes of middle-class observers whose fears about the effects of new technology needed to be allayed (p. 20).
29. Barringer, pp. 21-81. The significance of Brown's painting is also addressed by Curtis; Johnson; Klingender, pp. 171-8; and Williams, p. 53.
30. In 1858, the vestry of St John at Hampstead had planned extensive sewer works in conjunction with the commencement of the building of the northern high-level sewer, the head of which was not far away. It appears that, before 1858, there was a sewer in existence in The Mount. What Brown depicts in *Work* is excavation of the sewer at the point where it is needed to fall sharply into what would now be called a backdrop, The Mount (from where the sewer is coming) being much higher than the highway to the east (where the sewer is going); hence the reason for the extensive depth of the trench. I am grateful to Robin Winters at Thames Water plc for providing this information.
31. *Illustrated London News*, 18 March 1865, p. 266, 'Exhibition of Mr. Madox Brown's *Work*'.
32. *Illustrated London News*, 26 March 1843, pp. 226-8, 'Thames Tunnel, Opened 25 March 1843'.
33. Ibid., p. 227.
34. Barringer, p. 32.
35. *Illustrated Times*, 24 April 1869, p. 265, 'The Shield used in the Construction of the Tower Subway' and 'The Subway under the Thames, Near the Tower, Now in Course of Construction'; 18 September 1869, frontispiece, 'Construction of Barlow's Tunnel Under the Thames: Advancing the Shield'; and 25 September 1869, p. 196, 'Construction of Barlow's Tower Subway: Removing the Clay'.
36. Barringer, p. 49.
37. *Weekly Times*, 13 October 1861, p. 4.
38. *Morning Post*, 9 October 1861, p. 5, 'The London Sewerage'.
39. *Weekly Times*, 13 October 1861, p. 4.
40. *Building News*, 11 October 1861, p. 815, 'Visit of the Metropolitan Vestries to the Main Drainage Works'.
41. *Morning Post*, 9 October 1861, p. 5.
42. *Globe and Traveller*, 9 October 1861, p. 3, 'The Main Drainage of London'.
43. *Daily Telegraph*, 10 October 1861, p. 3, 'Metropolitan Main Drainage'.
44. *Morning Post*, 9 October 1861, p. 5.
45. *Daily Telegraph*, 10 October 1861, p. 3.
46. *Daily News*, 9 October 1861, p. 5, 'Metropolitan Main Drainage: the Board of Works'.
47. *Observer*, 13 October 1861, p. 6, 'The Main Drainage of the Metropolis: Inspection of the Works by the Metropolitan Board and Representative Vestries'.
48. Reprinted in Jennings and Madge, pp. 168-9.
49. Ibid., p. 168.
50. Ibid., p. 169.
51. *Illustrated London News*, Supplement, 30 November 1861, pp. 551, 555-6.
52. Sinnema, pp. 126-9.
53. One striking example is the engraving on the front page of the *Penny Illustrated Paper*, 11 November 1865, 'The Disastrous Explosion of the London Company's Gasholder, at Nine Elms', which depicts, in close-up detail, the moment when the gasholder exploded. By contrast, the engraving 'Ruins of the Gasworks at Nine-Elms after the Explosion',

published in the *Illustrated London News*, 11 November 1865, p. 465, shows the aftermath of the explosion from a more respectable distance.

54. *Penny Illustrated* Paper, 7 June 1862, p. 365, 'Terrific Explosion of Gas at Shoreditch'.
55. Ibid.
56. *Illustrated Weekly News*, 31 May 1862, p. 531, 'Terrible Explosion at Shoreditch'.
57. *Illustrated Times*, 28 June 1862, p. 139, 'The Bursting of the Fleet Ditch'.
58. *Penny Illustrated Paper*, 28 June 1862, p. 413, 'The Bursting of the Fleet Ditch, Clerkenwell'.
59. *Illustrated London News*, 28 June 1862, p. 647, 'Bursting of the Fleet Ditch'.
60. *Illustrated Times*, 28 June 1862, p. 139.
61. *Penny Illustrated Paper*, 28 June 1862, p. 413.
62. *Illustrated Times*, 28 June 1862, p. 139.
63. *Penny Illustrated Paper*, 28 June 1862, p. 413.
64. *Illustrated Times*, 28 June 1862, p. 139.
65. Anderson discusses the gradual displacement of the educational focus of illustrated periodicals like the *Penny Magazine* in the 1850s by a more sensationalist approach (pp. 98-115, 197). It is notable that the *Illustrated London News* generally resisted this trend until the 1870s, when it first began to publish serialised fiction – a key component in the success of the cheaper illustrated newspapers in the 1850s and 1860s (Houfe, p. 74).
66. Williams, p. 64.

Chapter 5

1. Pevsner and Radcliffe, p. 275.
2. Curl, (2007), p. 455.
3. Wilson, p. 36.
4. Avery and Stamp, p. 83.
5. Dixon and Muthesius, p. 116.
6. Curl (1973), p. 95.
7. Cherry, O'Brien and Pevsner, p. 230.
8. 'Obituary: Mr. C.H. Driver', *Journal of the Royal Institute of British Architects*, 7 (10 November 1900), p. 22; *Builder*, 10 November 1900, pp. 423-4; and *Minutes of Proceedings of the Institution of Civil Engineers* 143 (1901), pp. 341-2. Driver, who lived at 9 Gauden Road in Clapham in the latter part of his life, was survived by his wife Caroline Kempster (b.1837) and his six children, Margaret (b.1862), Charles (b.1863), Harry (b.1865), Amy (b.1867), Walter (b.1870) and Ernest (b.1872). I am grateful to Michael Dunmow for this information obtained from the 1881 census records.
9. *Builder*, 10 November 1900, p. 423.
10. *Minutes of Proceedings of the Institution of Civil Engineers*, 143 (1901), p. 341.
11. Institution of Civil Engineers, *Candidate Circulars*, Session 15, pp. 6-7, no. 238.
12. Felstead, Franklin, and Pinfield, pp. 562-3.
13. Piggott, p. 128. For Driver's opinions on the design of aquaria, see *Builder*, 4 March 1876, pp. 212-13; and 11 March 1876, pp. 243-4, 'Aquaria and their Construction'.
14. See *Architect*, 3 April 1869, pp. 179-80. On the Santiago market also see Higgs, p. 85. Photographs of the market held at the British Architectural Library in London show that the building was prefabricated in England and reassembled in Santiago from 1870 to 1872.
15. Cherry, Nairn, and Pevsner, p. 196.
16. Pevsner and Sherwood, p. 438 and *Builder*, 10 August 1872, pp. 625-6.
17. Pevsner and Cherry, p. 444.
18. *Builder*, 6 November 1858, pp. 746-7, 'The Ellesmere Memorial, Worsley, Lancashire'.
19. *Minutes of Proceedings of the Institution of Civil Engineers*, 143 (1901), p. 342.
20. Institution of Civil Engineers, *Candidate Circulars*, session 15, p. 6, no. 238.
21. LMA, MCS/482, 16 August 1849, p. 219.
22. Biddle, pp. 65, 265, 275, 279-80. Many of these stations, all of which were built in 1857, still survive: Southill and Cardington stations in Bedfordshire are now private houses; Oakley station, and its adjoining goods shed and office, is a commercial property;

Wellingborough and its adjoining buildings are still in use as a station and have been recently restored; Kettering is also still in use as a station and retains its original ironwork by Driver but the building is much altered; Glendon, Rushton and Desborough stations in Northamptonshire are private houses.

23. Biddle, pp. 106-7, and *Illustrated London News*, 24 August 1867, pp. 201-2, 'The Leatherhead and Dorking Railway'. The overtly ornamental stations on this section of the Leatherhead to Horsham line (Leatherhead, West Humble, and Dorking) were designed to appease the local landowner, Thomas Grissell, who did not want the new railway to spoil the natural beauty of this part of the North Downs.
24. Biddle, p. 81. Portsmouth station, built in 1866 and still extant, was a shared terminus for the London, Brighton and South Coast Railway and the London and South Western Railway.
25. Ibid., p. 103. Tunbridge Wells station (west) was built in 1866 and is still extant.
26. Ibid., pp. 9-10, 17, 20, 27, 40. Built 1865-7, the extant artefacts are the stations at Grosvenor Road, Battersea Park, Denmark Hill and Peckham Rye, and the London Bridge viaduct and train shed.
27. Graves, vol. 2, p. 370, entry 1211.
28. MBW Minutes, 3 November 1865, p. 1177, s. 4. Driver was paid £35 for these watercolours. One is located in the lecture room at the Crossness pumping station, while the other is probably in the office of the chief executive of Thames Water plc.
29. Graham and Waters, p. 95.
30. *Builder*, 28 January 1882, pp. 100-1, 'The Central Station of the Vienna Circular Elevated Railway', and 'Vienna City Railways Design for Central Station: Mr Joseph Fogerty, C.E., Architect' (engraving based on Driver's watercolour).
31. Bazalgette (1865), pp. 122-3, 134-5, 139-43.
32. See Smith (1972), p. 325 for details of important surviving examples of these.
33. Rolt, pp. 58-60.
34. Bazalgette (1868), pp. 4-9.
35. MBW/2320, 10 June 1868, p. 5.
36. MBW Minutes, 4 December 1868, p. 1304, s. 10; 23 September 1864, p. 905, s. 1; and 25 June 1869, pp. 753-5, s. 8. The enormous cost of Abbey Mills may be compared with the Horton Infirmary: around £6,200 (*Builder*, 10 August 1872, p. 625).
37. Bazalgette (1868), p. 7.
38. TWA, Works-as-executed Collection: Abbey Mills Pumping Station, Contract Drawings, Buildings &c., 1865.
39. MBW Minutes, 10 May 1867, p. 550, s. 4, payments 2150 and 2151.
40. Ibid., 8 July 1864, p. 737, s. 6.
41. Ibid., 3 November 1865, p. 1177, s. 4.
42. Graves, p. 150, entries 808 and 809.
43. *Building News*, 5 May 1865, p. 313, 'Exhibition of Architectural Drawings'. The article states: '[w]e should like to know who supplies Mr. Bazalgette with architectural ideas; we fear that, though put forward as his own, the merit of them is due to another'.
44. MBW Minutes, 4 August 1865, p. 984, s. 30. For a proof copy of this draft specification see TWA, Main Drainage Metropolis, Abbey Mills Pumping Station Buildings – Specification of and for the Engine & Boiler Houses, Coal Vaults, Dwelling Houses, River Wall, Sewers, &c., 1865.
45. Ibid., 20 April 1866, pp. 489-90, s. 4.
46. LMA, MBW/966: Main Drainage Committee, Minutes of Proceedings, pp. 100-2, 127-9; and MBW Minutes, 21 October 1859, p. 786, s. 10. Hawksley also assisted Bazalgette with his final report on the main drainage system presented on 6 April 1858. The design of the Deptford pumping station closely mirrors that of Hawksley's other architectural work, in particular the Humbledon pumping station (*c.*1852) in Seaforth Road, Sunderland.
47. Crook, p. 47.
48. Ruskin (1849), pp. 8-214.
49. M.W. Brooks, pp. 17-74.

50. Ruskin (1849), pp. 119, 121-2. Ruskin singles out the decoration of railways stations for his explicit condemnation.
51. O. Jones (1856), pp. 14-51.
52. Scott, pp. x, 189, 200.
53. Curl (2007), 457-8.
54. Ruskin (1849), pp. 137-43.
55. Crook, pp. 76-7.
56. Many examples can be found in the City of London, such as Frederick Jameson's offices at 103 Cannon Street (1860), George Atchinson's at 59-61 Mark Lane (1864) and Thomas Chatfeild Clarke's at 25 Throgmorton Street (1869).
57. Throughout 1851 issues of the *Builder* featured details of Venetian palaces from the 14th to the 16th centuries.
58. Crook, pp. 93-7, 145, 147.
59. See Dean. Driver assisted the engineer Frederick Banister in the design of the new train shed for the London, Brighton and South Coast Railway as well as the extensive viaduct to the south of the station.
60. *Builder*, 15 March 1851, p. 171, 'The Palace Dei Pergoli Intagliati, Venice' (engraving).
61. See Macfarlane, pp. 10, 37-8. Macfarlane's catalogues from this period marketed this specific cast-iron spiral drainpipe and the company applied it to its own foundry buildings in Glasgow and London in the early 1860s.
62. Ruskin (1851-53), vol. 1 (1851), pp. 128-9.
63. Scott, p. 262; and Street, p. 276.
64. Street, pp. 256, 268.
65. Such as the Palazzo Medici (1444-59) in Florence or the Palazzo Vendramin Calergi (*c.*1500-09) in Venice.
66. This window type was used in the interior of Paddington station (1852-54) and the Hop Exchange (1866) in Southwark Street.
67. Saunders, p. 310.
68. Curl (2007), p. 457.
69. Wilson, p. 36.
70. Avery and Stamp, p. 83.
71. On the University Museum, Oxford, see Curl (2007), pp. 223-8; Acland and Ruskin; Blau, p. 48-81; M.W. Brooks, pp. 113, 117-34; Crook, pp. 78-9; and Dixon and Muthesius, pp. 159-60.
72. For a comprehensive description of the Cistercian Abbey of St Mary, Stratford, see Barber et al. Remains of the Abbey were uncovered between 1991 and 1993, during the excavations for the Jubilee Line Extension.
73. Walford, vol. 1, p. 507.
74. TWA, Specification of and for the Engine & Boiler Houses, Coal Vaults, Dwelling Houses, River Wall, Sewers, &c., 1865, p. 31, clause 71 and 1866, p. 24, clause 71.
75. Exemplified by the 'pavilion principle' introduced in the 1850s by the architect Henry Currey (1820-1900), one of Driver's nominees for RIBA membership in 1872. Currey designed hospitals, such as St Thomas's (1868-71), according to this principle, which connected sanitation and ventilation in the form of airy pavilions (Felstead, Franklin and Pinfield, vol. 1, pp. 228-9).
76. The chimneys were removed in 1940, reputedly to prevent their use as navigation aids by German bombers, but more likely for the safety of the pumping station in the event of an air attack.
77. Bazalgette (1868), p. 7.
78. Ruskin (1849), p. 76.
79. Hitchcock, vol. 1, p. 515.
80. *Builder*, 25 April 1857, pp. 230-1, 'Suggestions for Furnace Chimneys', and 'Designs for Furnace Chimney-shafts: from Sketches by Mr. R. Rawlinson' (engraving).
81. *Builder*, 26 September 1868, p. 719, 'Cost of Abbey Mills Pumping Station'.
82. *Builder*, 6 November 1858, p. 746, 'The Ellesmere Memorial, Worsley, Lancashire' and p.

747, 'The Proposed Ellesmere Memorial, Worsley, Lancashire: Messrs. Driver and Webber, Architects' (engraving).
83. See Network Rail, Plan and Records Store, London, Drawings No. 2584E, 02 & 03, Dorking Station, Front and End Elevation. Dorking Station was built in 1867 but was demolished and rebuilt in the 1980s.
84. Cherry and Pevsner (1973), p. 444. It is uncertain how Driver became involved in this project, given his lack of experience in church restoration. The sponsor of this project was Mary Ann Horton (1789-1869), the 'Lady of the Manor' in the nearby village of Middleton Cheney. She also employed Driver to design the Horton Hospital in 1869, probably on the basis of his work at St Mary's, but it is uncertain as to how she knew of Driver's recently established London practice. One possible answer may lie in social connections: Driver was a prominent mason in his later life.
85. Ruskin (1849), p. 54.
86. Ibid., p. 57.
87. Driver (1879), p. 5.
88. Driver (1874), pp. 10-11.
89. Driver (1875), pp. 165-83.
90. Ibid., p. 166.
91. Ibid., p. 167.
92. Ibid., p. 168.
93. Ibid., p. 180.
94. Ibid., p. 179.
95. Ibid., p. 180.
96. Ibid., pp. 178-9.
97. George Aitchison, 'On Iron as a Building Material', *RIBA Transactions,* 1st series, 14 (1864), pp. 97-107.
98. O. Jones (1862), p. 22.
99. For a discussion of Jones's attitude towards iron, see Darby and Van Zanten.
100. See Wyatt (1850), and also Wyatt (1852). On Wyatt's contribution to Paddington station and his working relationship with Brunel see Saint, pp. 113-18.
101. Fergusson (1862), p. 474.
102. Driver (1874), pp. 5-6.
103. Ibid., pp. 7, 9.
104. Ibid., p. 9.
105. Driver (1875), p. 179.
106. Ibid., p. 180.
107. On octagonal markets see Carls and Schmeichen, pp. 254-5, 293; on octagonal glasshouses, Kohlmaier and von Sartory, pp. 180-3, 234-6, 284-6, 301-2.
108. Curl (2007), p. 457.
109. Bullen.
110. Ruskin (1851-53), vol. 2 (1852), p. 151.
111. Ruskin (1851-53), vol. 1 (1851), pp. 13-17, 21.
112. Wyatt and Waring, p. 15.
113. Ruskin (1851-53), vol. 1 (1851), pp. 274, 278-81.
114. P. Thompson, p. 288.
115. Muthesius (1972), p. 197.
116. Biddle, pp. 59-60; and Humber, pp. 3-15.
117. *Builder*, 15 August 1868, p. 603, 'New Lamp Lately Erected in Holborn' and p. 604, 'Iron Lamp, Recently set up in Holborn: Designed by Mr. Chas. H. Driver' (engraving). At some point the lamp was disassembled; parts of it were used to form the lamp now situated outside the main entrance of the Law Courts in the Strand.
118. Ruskin (1851-53), vol. 1 (1851), p. 245.
119. On Ruskin's design for the spandrel see Acland and Ruskin, p. 88; Blau, pp. 61-2; and Higgs, pp. 159-60. The revised design was illustrated in the *Builder*, 7 July 1855, p. 318.
120. Driver (1875), p. 173.
121. Darby and Van Zanten, pp. 57-65.

122. O. Jones (1856), p. 5, 'Proposition 8'.
123. Driver (19 December 1879), pp. 6-7.
124. See Halliday, pp. 108-23.
125. Driver (1875), pp. 175, 182.
126. On the buildings for the International Exhibition of 1862 see Anon. (1862). On the Midland Grand Hotel, see Simmons and Thorne, pp. 56-89.
127. See for example Macfarlane. Included in this catalogue are several of Driver's own designs: a lamp erected in Holborn, London in 1868 (pp. 34-5); iron railings used at Battersea Park station in 1866 (pp. 50-1); cresting used at Portsmouth station in 1866 (p. 7); and columns used at Battersea Park and Denmark Hill stations in 1866 and at Leatherhead station in 1867 (pp. 50-1).
128. See Carls and Schmeichen, pp. 51-3.

Chapter 6

1. Hollingshead, p. 390.
2. A point made by the *East London Observer*, 8 August 1868, p. 5, 'Visitation of Abbey Mills Pumping Station'.
3. On the planning of the ceremony at Crossness see LMA, MBW/975: Sub-committee on Opening of the Main Drainage Works, Minutes of Proceedings, pp. 445-71. On the planning of the Abbey Mills ceremony see MBW Minutes, 12 June 1868, pp. 761-2, s. 8; 3 July 1868, p. 879, s. 81; 17 July, p. 918, s. 6; 24 July 1868, p. 956, s. 1; and 7 August 1868, p. 1041, s. 1.
4. MBW Minutes, 3 July 1868, p. 879, s. 81.
5. *Illustrated London News*, 20 December 1851, pp. 725-6, 'Opening of the Croydon Water Works'.
6. *Illustrated London News*, 17 January 1863, pp. 73-4, 'Opening of the Metropolitan Railway'.
7. Bourne, pp. 243, 261-2, 270-1.
8. Ibid., p. 243.
9. Bazalgette (1865), pp. 280-314.
10. *The Times*, 4 April 1865, p. 14, 'The Main Drainage of the Metropolis'; *Standard*, 4 April 1865, p. 6, 'The Southern Outfall'; and *Morning Post*, 5 April 1865, p. 5, 'The Main Drainage System'.
11. *Illustrated London News*, 8 April 1865, p. 335, 'The Metropolitan Main Drainage Southern Outfall at Crossness'.
12. LMA, MBW/2511: Southern Outfall Works, Buildings, 1862, Contract Drawing no. 1, general plan.
13. Bazalgette (1868).
14. *The Times*, 31 July 1868, p. 12, 'The Thames Embankment'.
15. *Illustrated London News*, 15 August 1868, p. 162, 'The Metropolitan Main Drainage'.
16. *Daily Telegraph*, 5 April 1865, p. 2, 'Opening of the Main Drainage by the Prince of Wales'.
17. *Morning Post*, 5 April 1865, p. 5, 'The Main Drainage System'; *Daily Telegraph*, 5 April 1865, p. 2; *Standard*, 4 April 1865, p. 5, 'The Southern Outfall'; and *City Press*, 8 April 1865, p. 9, 'Completion and Opening of the Main Drainage Works at Crossness'.
18. *Daily Telegraph*, 5 April 1865, p. 2.
19. *Marylebone Mercury*, 8 August 1868, p. 2, 'The Abbey Mills Pumping Station'.
20. *The Times*, 4 April 1865, p. 14, 'The Main Drainage of the Metropolis' and 31 July 1868, p. 12, 'The Thames Embankment'; *Observer*, 9 April 1865, p. 5, 'Opening of the Southern Outfall of the Main Drainage Works'; *Standard*, 31 July 1868, p. 3, 'Opening of the Thames Embankment Footway'; and *City Press*, 8 August 1868, p. 3, 'The Abbey Mills Pumping Station of the Main Drainage Works: Visit of the Corporation'.
21. *Morning Star*, 5 April 1865, p. 5.
22. *The Times*, 31 July 1868, p. 12.
23. *The Times*, 4 April 1865, p. 14.
24. *Standard*, 4 April 1865, p. 5.
25. See Reid, pp. 35-6.

26. *Observer*, 9 April 1865, p. 5.
27. *The Times*, 5 April 1865, p. 5, 'Opening of the Main Drainage'.
28. *Daily News*, 5 April 1865, p. 5, 'Opening of the Metropolitan Main Drainage Works by the Prince of Wales'.
29. *The Times*, 31 July 1868, p. 12.
30. *Observer*, 2 August 1868, p. 3, 'Thames Embankment and Abbey Mills Pumping Station'.
31. *Standard*, 31 July 1868, p. 3.
32. *Daily Telegraph*, 5 April 1865, p. 2.
33. *Daily News*, 5 April 1865, p. 5; and *The Times*, 5 April 1865, p. 5.
34. *Daily Telegraph*, 5 April 1865, p. 2.
35. *Morning Star*, 5 April 1865, p. 5; *Standard*, 5 April 1865, p. 3; and *Daily Telegraph*, 5 April 1865, p. 2.
36. *Daily Telegraph*, 5 April 1865, p. 2.
37. *Illustrated London News*, 15 April 1865, p. 341, 'Opening the Metropolitan Main-drainage Works at Crossness: the Prince of Wales Starting the Engines' (engraving), p. 342, 'The Prince of Wales at the Metropolitan Drainage Works', p. 344, 'Mr. Bazalgette Explaining the Main-drainage Plans' and 'The Luncheon in the Workshop at Crossness' (engravings), p. 345, 'The Prince of Wales Opening the Metropolitan Main-drainage Works at Crossness: Interior of the Engine-house' (engraving), and p. 348, 'The Prince of Wales Opening the Metropolitan Main-drainage Works at Crossness: the Underground Reservoir Illuminated' (engraving).
38. *City Press*, 8 April 1865, p. 9.
39. *Morning Star*, 5 April 1865, p. 5.
40. *Daily Telegraph*, 5 April 1865, p. 2.
41. Ibid.
42. *Daily Telegraph*, 10 January 1863, p. 3, 'Opening of the Metropolitan Railway'.
43. Hugo, p. 1071.
44. *Daily Telegraph*, 5 April 1865, p. 2; and *Daily News*, 5 April 1865, p. 5.
45. Ibid.
46. *The Times*, 31 July 1868, p. 12.
47. *East London Observer*, 8 August 1868, p. 5.
48. *South London Press*, 8 August 1868, p. 6, 'A Palace of Filth: Visit of the South London Vestries to Abbey Mills'.
49. *East London Observer*, 8 August 1868, p. 5.
50. MBW/975, p. 446.
51. *Morning Post*, 5 April 1865, p. 6.
52. *Morning Star*, 5 April 1865, p. 5.
53. *Daily Telegraph*, 5 April 1865, p. 2.
54. *Standard*, 5 April 1865, p. 3.
55. *The Times*, 4 April 1865, p. 14.
56. *Morning Post*, 5 April 1865, p. 5.
57. *The Times*, 4 April 1865, p. 14.
58. Booth; Clarke; and Wiggins. Jasper Rodgers, honorary engineer to the Sanitary Association of Dublin, published an equally inflammatory pamphlet in 1858 titled *Fact and Fallacies of the Sewerage System of London and Other Large Towns*. Rodgers' account, mirroring Chadwick's views, proposes an end to all existing sewers, replacing them with small-bore airtight pipes to collect and recycle urban sewage.
59. Booth, pp. 17-18; Clarke, pp. 24-6; and Wiggins, p. 11.
60. Clarke, p. 18.
61. Ibid., pp. 22-4, 27-8, 32.
62. Booth, p. 6.
63. Wiggins, pp. 23-4.
64. Booth, pp. 7-8.
65. Ibid., p. 20.
66. Mayhew, vol. 2, pp. 388-425.

67. Allen, p. 34. For the account of the nightmen see Mayhew, vol. 2, pp. 451-2.
68. Pike (2005), p. 200. For Mayhew's descriptions of the sewer hunters and mud-larks see Mayhew, vol. 2, pp. 150-8.
69. Mayhew, vol. 2, plate between pp. 334-5, 'London Nightmen'; plate between pp. 370-1, 'The Rat-catchers of the Sewers'; and plate between pp. 388-9, 'The Sewer-hunter'. All the plates are supposedly based on daguerreotypes.
70. Hollingshead was a prolific London-based journalist: in 1857 he worked for Dickens on *Household Words* while later contributing to the *Morning Post*, the *Leader*, the *London Review* and *Good Words*; From 1863 to 1868 he was drama critic for the *Daily News* and was director of the Alhambra Theatre from 1865 to 1868.
71. Hollingshead, p. 2.
72. Ibid., pp. 1, 4.
73. Ibid., pp. 87-8, 98-9.
74. Nead., p. 6.
75. *East London Observer*, 8 August 1868, p. 5.
76. *Daily Telegraph*, 31 July 1868, p. 2, 'Opening of the Thames Embankment Footway'.

Postscript

1. Drabble, p. 116.
2. Carey, pp. 319-21.

Newspaper articles and engravings consulted

Illustrated newspapers

Architect

3 April 1869, pp. 179-80, 'New Market for Santiago' and p. 180, 'Plan of the New Market for Santiago' (engraving) and 'Santiago Market' (2 engravings)

Builder

3 June 1843, pp. 207-9, 'Mr. Chadwick and the Surveyors'
3 June 1848, p. 274, 'Metropolitan Survey'
17 June 1848, pp. 291-2, 'What are the Crows' Nests for? Ordnance Survey of the Metropolis'
15 March 1851, p. 170, 'Il Palazzo Dei Pergoli Intagliati, Venice' and 'Details of Palace Dei Pergoli Intagliati' (engraving)
15 March 1851, p. 171, 'The Palace Dei Pergoli Intagliati, Venice' (engraving)
24 May 1851, p. 330, 'Casa Visetti, Venice' and 'Details of Casa Visetti, Venice' (engraving), and p. 331, 'Casa Visetti, Venice' (engraving)
27 December 1851, p. 815, 'Capitals: Ducal Palace, Venice' and 'Capitals: Ducal Palace, Venice' (engraving)
3 June 1854, p. 290, 'The Paddington Station of the Great Western Railway' and p. 291, 'The Paddington Station of the Great Western Railway: the Joint Design of Mr. I. K. Brunel and Mr. M. D. Wyatt' (engraving)
17 June 1854, p. 322, 'The Paddington Station of the Great Western Railway' and p. 323, 'Entrance to Offices, Paddington Station' (engraving)
23 September 1854, p. 498, 'The Byzantine Court in the Crystal Palace, Sydenham' and p. 499, 'The Byzantine Court in the Crystal Palace, Sydenham' (engraving)
25 April 1857, p. 230, 'Suggestions for Furnace Chimneys' and p. 231, 'Designs for Furnace Chimney-shafts: from Sketches by Mr. R. Rawlinson' (engraving)
2 May 1857, p. 243, 'Worthing Water-tower and Engine-house' and 'Worthing Water-tower and Engine-house: Mr. Rawlinson, engineer' (engraving)
6 November 1858, p. 746, 'The Ellesmere Memorial, Worsley, Lancashire' and p. 747, 'The Proposed Ellesmere Memorial, Worsley, Lancashire: Messrs. Driver and Webber, Architects' (engraving)

4 June 1859, pp. 371-2, leader
18 February 1860, pp. 97-9, 'The Metropolitan Main-drainage: the Deptford Pumping Station and Machinery'
3 March 1860, p. 130, 'The Metropolitan Main-drainage: the Deptford Pumping Station and Machinery'
31 March 1860, pp. 193-5, 'Progress of the Metropolitan Main-drainage'
6 October 1860, pp. 633-4, 'The Metropolitan Main-sewerage: the Northern Outfall Sewer'
27 October, 1860, p. 691, 'The Great Northern Outfall Sewer'
15 December 1860, p. 806, 'The Progress of the Main Drainage Works'
25 January 1862, pp. 53-5, 'Progress of the Metropolitan Main Drainage: the Works North of the River'
22 March 1862, p. 206, 'Drinking Fountain, Kennington Park: Mr. C. H. Driver, Architect' (engraving) and 'Drinking Fountain, Kennington Park'
15 February 1862, p. 114, 'Portable Steam-crane' and 'Taylor's Portable Steam-crane: Metropolitan Drainage Works' (engraving)
19 September 1863, pp. 676-7, 'The Main Drainage and Suburban Gardens'
16 July 1864, pp. 521-2, 'The Northern Low-level Sewer'
14 January 1865, p. 29, 'Lamp and Ventilating Shaft Erected over the Subway, Southwark Street, London: Executed by Messrs. Walter Macfarlane & Co., Under the Direction of Mr. Bazalgette' (engraving) and p. 31, 'Lamp Standard and Ventilating Shaft, Southwark Street, London'
19 August 1865, p. 591, 'Engine-house, Crossness: Outfall of the Southern Metropolitan Sewerage. Erected Under the Direction of Mr. Bazalgette, Engineer to the Metropolitan Board of Works' (engraving) and p. 593, ' The Engine-house at the Outfall of the Southern Metropolitan Sewerage'
25 March 1865, pp. 206-8, 'The Drainage of London, Institution of Civil Engineers'
23 December 1865, p. 908, 'The Main Drainage of London'
13 January 1866, 'Metropolitan Board of Works: Progress of the Main Drainage and Thames Embankment Works'
1 August, 1868, pp. 569-70, 'New Works of the Board of Works'
15 August 1868, p. 603, 'New Lamp Lately Erected in Holborn' and p. 604, 'Iron Lamp, Recently set up in Holborn: Designed by Mr. Chas. H. Driver' (engraving)
26 September 1868, p. 719, 'Cost of Abbey Mills Pumping Station'
10 August 1872, p. 625, 'The Horton Infirmary, Banbury' and 'The Horton Infirmary, Banbury: Mr. Charles H. Driver, Architect' (engraving), and p. 626, 'The Horton Infirmary, Banbury: Plan' (engraving)
21 September 1872, p. 746, 'The Vienna Universal Exhibition in 1872' and 'Vienna Universal Exhibition: Block Plan of the Building and its Surroundings' (engraving), and p. 747, 'Vienna Universal Exhibition: View of the Central Rotunda' (engraving)
4 March 1876, pp. 212-13 and 11 March, pp. 243-4, 'Aquaria and their Construction'
28 January 1882, pp. 100-1, 'The Central Station of the Vienna Circular

Elevated Railway' and 'Vienna City Railways: Design for Central Station: Mr Joseph Fogerty, C.E., Architect' (engraving)
10 November 1900, pp. 423-4, 'Obituary: Mr. C. H. Driver'

Building News
11 October 1861, pp. 815-16, 'Visit of the Metropolitan Vestries to the Main Drainage Works'
20 March 1863, p. 219, 'Details from the Grosvenor Hotel, Pimlico: J.T. Knowles, Esq., Architect' (engraving)
5 May 1865, p. 313, 'Exhibition of Architectural Drawings'
19 May 1865, p. 351, 'Exhibition of Architecture'
28 August 1868, pp. 589-90, 'The Thames Embankment'
7 February 1879, pp. 140-1, 'Engineering and Art'

Civil Engineer and Architect's Journal
Vol. 15, 1852, p. 160, 'Mr. Frank Forster'

Engineer
11 October 1861, p. 213, 'The Main Drainage of London'
7 April 1865, p. 211, 'Opening of the Main Drainage'
30 August 1867, p. 176, 'Abbey Mills Pumping Station, Metropolitan Main Drainage. Ground Plan of Mains and Details of Pumps &c.' (engraving), pp. 178-9, 'Abbey Mills Pumping Station – Metropolitan Main Drainage, Northern Outfall', and p. 180, 'Pumping Engine, Abbey Mills, Metropolitan Main Drainage. Mr Bazalgette, Engineer' (engraving)
6 September 1867, p. 203, 'Abbey Mills Pumping Station – Metropolitan Main Drainage, Northern Outfall' and 'Sectional Elevation and Plan of Boiler' (engraving) and p. 206, 'Abbey Mills Pumping Station – Front Elevation of Engine House. Mr J. W. Bazalgette, Engineer' (engraving)
13 September 1867, p. 226, 'Abbey Mills Pumping Station – Half-sectional Elevation Through the Boiler House' (engraving)
20 September 1867, p. 248, 'Abbey Mills Pumping Station – Details of Roofing' (engraving) and p. 250, 'Abbey Mills Pumping Station'
27 September 1867, p. 270, 'Abbey Mills Pumping Station – Central Pillars and Entablatures' (engraving), p. 271, 'Abbey Mills Pumping Station – Filth Hoist, Penstock, and General Plan' (engraving), pp. 272-3, 'Abbey Mills Pumping Station', and pp. 278-9, 'Sectional Elevation of Engine House, Abbey Mills Pumping Station, Metropolitan Main Drainage. Mr. J. W. Bazalgette, Engineer' (engraving)
31 July 1868, p. 85, 'Opening of the Thames Embankment Footway, and Inspection of the Abbey Mills Pumping Station'

Illustrated London News
26 March 1843, pp. 226-8, 'Thames Tunnel, Opened 25 March 1843' and 'The Company's Medallion of Sir Isambert [sic.] Brunel, Supported by One of the Tunnel Excavators' (engraving)

27 May 1843, p. 347, 'Our First Anniversary'
8 July 1843, 'Preface to Vol. II.'
4 October 1845, p. 213, 'The Fleet-street Sewer', and 'The Fleet-street Sewer' and 'Fleet-street: Deepening the Sewer' (engravings)
22 April 1848, p. 259, 'Survey of the Metropolis', and '"Crow's Nest"' on Westminster Abbey' and 'Cutting the "Black Mark"' (engravings)
24 June 1848, p. 414, 'Ordnance Survey of London and the Environs', and 'Observatory on the Cross of St. Paul's Cathedral', 'Interior of the Observatory' and 'Exterior of the Observatory' (engravings)
3 May 1851, pp. 343-4, 'The Great Exhibition' and pp. 348-9, 'The Opening of the Great Exhibition'
20 December 1851, p. 725, 'The Croydon Water Works: the Reservoir' (engraving) and p. 726, 'Opening of the Croydon Water Works'
23 May 1857, p. 506, 'Leicester and Hitchin Railway' and 'The Leicester and Hitchin Railway: Kettering Station' (engraving)
19 February 1859, p. 173, 'The Metropolitan Main Drainage' and 'Commencement of the Metropolitan High-level Sewer, near the Victoria Park' (engraving)
12 March 1859, p. 253-4, 'Mr. Bazalgette' (engraving)
30 April, 1859, p. 421, 'Commencement of the Main Drainage of the Metropolis' and 'Boring Operation, near Gray's-inn-lane, in Connection with the Main Drainage of the Metropolis' (engraving)
27 August 1859, p. 203, 'Main Drainage of the Metropolis' and 'Main Drainage of the Metropolis: Sectional View of the Tunnels from Wick-Lane, Near Old Ford, Bow, Looking Westward' (engraving)
Supplement, 30 November 1861, p. 551, 'London Main Drainage', and 'The Concrete Mills at Plaistow', 'Concrete Foundation for the Northern Outfall Tunnels' and 'Construction of the Concrete Embankment Across the Plaistow Marshes: Depositing the Concrete' (engravings); p. 554, 'Works at Barking Creek Outfall', 'Pumping-station at Deptford Creek', 'Driving a Tunnel at Peckham' and 'Bottom of a Shaft in the Southern High-level Sewer at Peckham' (engravings); p. 555, 'London Main Drainage', and 'The Penstock Chamber at Old Ford', 'Constructing the Invert for the Southern High-level Sewer' and 'Barrow-hoist on the Southern High-level Sewer at Peckham' (engravings); and p. 556, 'London Main Drainage', and 'Section of Main-drainage Work near Old Ford' (engraving)
28 June 1862, pp. 647, 'Bursting of the Fleet Ditch and Destruction of Part of the Metropolitan Railway: Scene of the Accident' (engraving) and p. 648, 'The Bursting of the Fleet Ditch'
17 January 1863, p. 73, 'Opening of the Metropolitan Railway: Banquet at the Farringdon-street Station' (engraving) and p. 74, 'Opening of the Metropolitan Railway'
21 May 1864, p. 493, 'The Metropolitan Main-drainage Works at Crossness' (engraving); pp. 501-2, 'The London Main-drainage Works, South Side' and p. 502, 'The London Main-drainage Works: Northern and Southern Outfalls' (engraving); and p. 504, 'The Metropolitan Main-drainage Works: Machinery for Lifting the Sewage' (engraving)

28 May 1864, p. 513, 'London Main-drainage Works: View and Section of the Outfall of the Northern Drainage at Barking Creek' (engraving) and pp. 512 and 514, 'London Main Drainage Works: North Side'
8 April 1865, p. 325, 'The Metropolitan Main Drainage: General View of the Southern Outfall Works at Crossness' (engraving); p. 328, 'The Metropolitan Main Drainage Works at Crossness: View in the Reservoir' (engraving); and p. 335, 'The Metropolitan Main Drainage Southern Outfall at Crossness' and 'Metropolitan Main Drainage: Plan of the Southern Outfall Works at Crossness' (engraving)
15 April 1865, p. 341, 'Opening the Metropolitan Main-drainage Works at Crossness: the Prince of Wales Starting the Engines' (engraving); p. 342, 'The Prince of Wales at the Metropolitan Drainage Works'; p. 344, 'Mr. Bazalgette Explaining the Main-drainage Plans' and 'The Luncheon in the Workshop at Crossness' (engravings); p. 345, 'The Prince of Wales Opening the Metropolitan Main-drainage Works at Crossness: Interior of the Engine-house' (engraving); and p. 348, 'The Prince of Wales Opening the Metropolitan Main-drainage Works at Crossness: the Underground Reservoir Illuminated' (engraving)
24 August 1867, pp. 201-2, 'The Leatherhead and Dorking Railway' and p. 201, 'Views on the Leatherhead and Dorking Railway' (engravings)
8 August 1868, p. 131, 'Opening of the Thames Northern Embankment to Foot Passengers: Embarkation at Temple Pier for Abbey Mills' (engraving) and pp. 133-34, 'Opening of the Thames Embankment'
15 August 1868, p. 161, 'General View of the Abbey Mills Pumping Station' and 'Interior of the Abbey Mills Pumping Station' (engravings) and p. 162, 'The Metropolitan Main Drainage'
29 January 1870, p. 128, 'The Sewers of Paris: "the Boat" and "the Wagon"' (engravings) and p. 129, 'The Paris Sewers'

Illustrated Times
14 December 1861, p. 377, 'Metropolitan Main Drainage Works - a Sketch at New Cross' and 'View Showing the Progress of the Works at New Cross' (engravings); pp. 381-2, 'The London Main-drainage System', and p. 381, 'Metropolitan Main-drainage Works: Sewer at Notting-Hill, Thirty-feet Below the Surface' (engraving)
28 June 1862, p. 139, 'The Bursting of the Fleet Ditch', and p. 140, 'The Bursting of the Fleet Sewer: View of the Damage Done to the Works of the Metropolitan Underground Railway' and 'Navvies Engaged in Turning the Course of the Fleet Ditch' (engravings)
22 August 1863, p. 121, 'The Metropolitan Drainage Outfall at Barking: Present State of the Works' (engraving) and p. 122, 'The Main Drainage and the Outfall at Barking Creek'
4 February 1865, p. 68, 'The Engine House at the Southern Outfall Sewer' and 'The Pumping-station at Crossness, of the Southern Outfall Sewer' (engraving)
8 April 1865, p. 211, 'Opening of the Main Drainage' and 'Ground Plan of the Metropolitan Main Drainage Works at Crossness' (engraving); p. 217, 'Banquet at Crossness, on the Occasion of Opening the Main Drainage System'

(engraving); and p. 220, 'The Metropolitan Sewers: General View of the Works at Crossness' and 'The Great Reservoirs and Culverts at Crossness' (engravings)
15 April 1865, frontispiece, 'The Prince of Wales Starting the Engines at the Main-drainage Works, Crossness' (engraving); p. 226, 'The Main-drainage Works at Crossness'; and p. 228, 'Interior of the Engine-house at the Main-drainage Works, Crossness' (engraving)
4 November 1865, p. 286, 'Terrible Explosion at Gasometer'
11 November 1865, p. 289, 'The Effects of the Late Explosion at Nine-Elms Gasworks' (engraving) and p. 290, 'The Gas Explosion at Nine Elms'
8 August 1868, frontispiece, 'Opening the Footway on the Thames Embankment' (engraving) and p. 82, 'The Thames Embankment'
22 August 1868, p. 120, 'London Main Drainage System: Abbey-Mills Pumping Station at Bow' (engraving) and p. 122, 'London Main Drainage: Abbey-Mills Pumping Station'

Illustrated Weekly News
31 May 1862, p. 531, 'Terrible Explosion at Shoreditch'
7 June 1862, frontispiece, 'The Scene of the Fearful Accident and Loss of Life in Shoreditch' (engraving) and p. 551, 'The Fearful Explosion in Shoreditch: Further Particulars'
28 June 1862, p. 599, 'The Accident to the Metropolitan Railway'

Penny Illustrated Paper
7 June 1862, p. 365, 'Terrific Explosion of Gas at Shoreditch' and 'The Frightful Explosion: Ruins of the Houses in Church-street' (engraving)
28 June 1862, p. 412, 'The Bursting of the Fleet Ditch: Navvies Constructing a Fresh Channel' (engraving) and p. 413, 'The Bursting of the Fleet Ditch'
11 November 1865, frontispiece, 'The Disastrous Explosion of the London Company's Gasholder, at Nine Elms' (engraving) and p. 370, 'The Fatal Gas Explosion at Nine Elms'
1 August 1868, p. 77, 'Opening of the Thames Embankment' and p. 78, 'Opening of the Thames Embankment: Design for Landing-stairs and Ornamental Gardens Between Hungerford and Waterloo Bridges' (engraving)
8 August 1868, pp. 84, 86, 'The Drainage of London' and p. 84, 'The London Main Drainage Works: Northern and Southern Outfalls' (engraving)

Non-illustrated newspapers

City Press
8 April 1865, p. 9, 'Completion and Opening of the Main Drainage Works at Crossness'
8 August 1868, p. 3, 'The Abbey Mills Pumping Station of the Main Drainage Works: Visit of the Corporation'

Daily News
9 October 1861, p. 5, 'Metropolitan Main Drainage: the Board of Works'

4 April 1865, p. 4, leader
5 April 1865, p. 5, 'Opening of the Metropolitan Main Drainage Works by the Prince of Wales'
31 July 1868, p. 5, 'Opening of the Thames Embankment and the Abbey Mills Pumping Station'

Daily Telegraph
10 October 1861, p. 3, 'Metropolitan Main Drainage'
10 January 1863, pp. 3-4, 'Opening of the Metropolitan Railway'
5 April 1865, p. 2, 'Opening of the Main Drainage by the Prince of Wales'
31 July 1868, p. 2, 'Opening of the Thames Embankment Footway'

East London Observer and Tower Hamlets Chronicle
8 August 1868, p. 5, 'Visitation of Abbey Mills Pumping Station'

Essex Times & Romford Telegraph
5 August 1868, p. 7, 'Opening of the Abbey Mills Pumping Station'
12 August 1868, p. 4, leader

Globe and Traveller
9 October 1861, p. 3, 'The Main Drainage of London'

Lloyd's Weekly London Newspaper
13 October 1861, p. 8, 'The Main Drainage Works'
9 April 1865, p. 5, 'Opening of the Main Drainage'

Marylebone Mercury
12 October 1861, p. 3, 'Metropolitan Main Drainage'
8 April 1865, p. 2, leader
8 August 1868, p. 2, 'The Abbey Mills Pumping Station'

Morning Herald
9 October 1861, p. 3, 'The New Metropolitan Sewer System'
10 October 1861, p. 6, 'Visit of the Vestries to the New Sewers'
4 April 1865, p. 6, 'The Southern Outfall'
5 April 1865, p. 3, 'Opening of the Southern Outfall Works'
31 July 1868, p. 6, 'Opening of the Thames Embankment Footway'

Morning Post
9 October 1861, p. 5, 'The London Sewerage'
10 October 1861, p. 4, leader
5 April 1865, pp. 5-6, 'The Main Drainage System'
31 July 1868, p. 6, 'The Thames Embankment and the Main Drainage'

Morning Star
4 April 1865, p. 2, 'The Metropolitan Main Drainage Works'
5 April 1865, p. 5, 'Opening of the Main Drainage Works by the Prince of

Wales'
31 July 1868, p. 6, 'The Opening of the Thames Embankment'

News of the World
13 October 1861, p. 3, 'Grand Inspection of the Main Drainage Works'
9 April 1865, p. 3, 'The Gigantic Main Drainage of the Metropolis'
2 August 1868, p. 4, 'Opening of the Thames Embankment'

Observer
14 April 1861, p. 5, 'The Metropolitan Great Drainage Works'
13 October 1861, pp. 5-6, 'The Main Drainage of the Metropolis: Inspection of the Works by the Metropolitan Board and Representative Vestries'
9 April 1865, p. 5, 'Opening of the Southern Outfall of the Main Drainage Works'
2 August 1868, p. 3, 'Thames Embankment and Abbey Mills Pumping Station'

Reynolds's Newspaper
13 October 1861, p. 14, 'The Main Drainage of London'
9 April, 1865, leader, 'Opening of the Main Drainage of the Metropolis by the Prince of Wales'

South London Journal
12 October 1861, p. 5, 'Metropolitan Main Drainage Works: Visit of the Metropolitan Board'
8 April 1865, p. 5, 'The Great Metropolitan System of Drainage: Opening of the Southern Works by the Prince of Wales'
1 August 1868, p. 7, 'The Thames Embankment and the Main Drainage System'

South London Press
8 April 1865, p. 8, 'Opening of the Main Drainage'
8 August 1868, p. 6, 'A Palace of Filth: Visit of the South London Vestries to Abbey Mills'

Standard
10 January 1863, p. 2, 'Railway Intelligence: the Metropolitan'
4 April 1865, pp. 5-6, 'The Southern Outfall'
5 April 1865, p. 3, 'Opening of the Southern Outfall Works'
31 July 1868, p. 3, 'Opening of the Thames Embankment Footway'

Stratford Express: East London & South Essex Advertiser
1 August 1868, p. 4, leader, p. 5, 'Opening of the Abbey Mills Pumping Station'

Sun
9 October 1861, p. 1, 'The London Sewerage'
4 April 1865, p. 1, 'The Main Drainage of the Metropolis'
5 April 1865, p. 2, 'Opening of the Main Drainage'
31 July 1868, p. 3, 'Opening of the Thames Embankment'

The Times
4 November 1848, p. 5, 'The Trigonometrical Survey of London'
1 February 1851, p. 3, 'Metropolitan Sewers Commission: Drainage of the Metropolis North of the Thames'
9 October 1861, p. 8, 'The Main Drainage of London'
10 October 1861, p. 8, leader
10 January 1863, p. 10, 'The Metropolitan Railway'
4 April 1865, p. 14, 'The Main Drainage of the Metropolis'
5 April 1865, p. 5, 'Opening of the Main Drainage'
31 July 1868, p. 12, 'The Thames Embankment'

Weekly Dispatch
13 October 1861, p. 14, 'The Main Drainage of London'
9 April 1865, p. 12, 'Opening of the Southern Outfall of the Main Drainage'
2 August 1868, p. 16, 'Opening of the Thames Embankment'

Weekly Times
13 October 1861, p. 4, 'The Great Main Drainage'

Bibliography

Manuscript and archival sources

Abbreviations

BAL British Architectural Library, London
ICE Institution of Civil Engineers, London
LMA London Metropolitan Archives, London
MBW Metropolitan Board of Works
MOL Museum of London, London
TWA Thames Water Archive, Abbey Mills Pumping Station, London

This list gives the location of manuscript collections consulted. Full references relating to individual documents explicitly referred to are given in the endnotes.

BAL, *Journal of the Royal Institute of British Architects*, vol. 7, 10 November 1900, p. 22

BAL, photographs depicting the Santiago market (*c.*1870)

BAL, *RIBA Nomination Papers*

ICE, *Candidate Circulars*, session 15, p. 6, no. 238

ICE, *Minutes of Proceedings of the Civil and Mechanical Engineers' Society*, 522, 19 December 1879, pp. 3-8

ICE, *Minutes of Proceedings of the Institution of Civil Engineers*, 143, 1901, pp. 341-2

LMA, MBW Reports: Main Drainage 1856-69, 20.101 MET

LMA, Ordnance Survey maps, SC/OS/LN, LCC/CE/MD, RM28/9

LMA, records of the MBW

LMA, records of the Metropolitan Commissions of Sewers (MCS)

Network Rail, Plan and Records Store, London, contract drawings for Dorking station, c.1866, drawings no. 2584E, 02 and 03

MOL, Henry Flather collection, 65 photographs depicting the construction of the Metropolitan Railway (1862)

TWA, Abbey Mills Pumping Station, Buildings, Specification of the Engine & Boiler Houses, Coal Vaults, Dwelling Houses, River Wall, Sewers, &c., (proof B), 1865; revised specification, June 1866; and Schedule of Prices for Works to be Executed by Mr. William Webster, 1866

TWA, alphabetical list of contracts under Metropolitan Commissions of Sewers and MBW

TWA, map made by the Holborn and Finsbury Sewer Commission (1846)

TWA, *Metropolis Main Drainage Progress Report*, October 1861

TWA, Ordnance Survey sheets (five feet to one mile) with later sewer overlays, *c.*1848

TWA, photographs depicting the building of Crossness and Deptford pumping stations and the northern outfall sewer (1860-5)

TWA, Works-as-executed collection, Main Drainage Metropolis, Abbey Mills Pumping Station, Buildings (1865)

TWA, Works-as-executed collection, Main Drainage Metropolis, Abbey Mills Pumping Station, Engines (1864)

Other unpublished sources

'Bricks and Water' exhibition held at the Kew Bridge Steam Museum, London, December 2003

Brookes, Jeff, 'When Abbey Mills was New', unpublished article, 2004

Guillery, Peter, '*Abbey Mills*': *Report by the Royal Commission on Historic Monuments of England*, unpublished report, Royal Commission on Historic Monuments of England, 1995

Higgs, M.S., 'Iron Architecture in Britain and America (1706-1880), with Special Reference to the Development of the Portable Building', unpublished doctoral thesis, University of Edinburgh, 1972

Hookes, Jo, '"Cathedrals of Sewage": a Discussion of the Architectural Style of London's Four Main Drainage Sewage Pumping Stations', unpublished master's thesis, Royal Holloway College, University of London, 1996

Pinkus, Rosa L., 'The Conceptual Development of London, 1850-1855', unpublished doctoral thesis, State University of New York at Buffalo, 1975

Other published sources

Ackroyd, Peter, *London, the Biography* (London, 2000)

Acland, Henry W., and John Ruskin, *The Oxford Museum* (London, 1996; original 1859)

Allen, Michelle E., *Cleansing the City: Sanitary Geographies in Victorian London* (Athens, Ohio, 2008)

Altick, Richard D., *The English Common Reader: a Social History of the Mass Reading Public, 1800-1900* (Columbus, Ohio, 1998)

Anderson, Patricia, *The Printed Image and the Transformation of Popular Culture, 1790-1860* (Oxford, 1991)

Anon., *Some Account of the Buildings Designed by Francis Fowke, Capt. R. E. for the International Exhibition of 1862, and Future Decennial Exhibitions of the Works of Art and Industry* (London, 1862)

International Exhibition, 1862: Official Catalogue of the Industrial Department (London, 1862)

The Crossness Engines (London, 2002)

Ashfield, Andrew, and Peter de Bolla (editors), *The Sublime: a Reader in British Eighteenth-century Aesthetic Theory* (Cambridge, 1996)

Barber, Bruno, et al., *The Cistercian Abbey of St Mary Stratford Langthorne, Essex* (London, 2004)

Barker, Theodore C., and Michael Robbins, *A History of London Transport: Passenger Travel and the Development of the Metropolis* (London, 1974)

Barnes, David S., 'Confronting Sensory Crisis in the Great Stinks of London and Paris', in William A. Cohen and Ryan Johnson (editors), *Filth: Dirt, Disgust and Modern Life* (Minneapolis and London, 2005)

Barringer, Tim, *Men at Work: Art and Labour in Victorian Britain* (New Haven and London, 2005)

Barton, Nicholas, *The Lost Rivers of London* (London, 1962)

Baynes, Ken, and Francis Pugh, *The Art of the Engineer* (Guildford, 1981)
Bazalgette, Joseph W., *A Short Descriptive Account of the Thames Embankment and of the Abbey Mills Pumping Station* (London, 1868)
'On the Main Drainage of London and the Interception of the Sewage from the River Thames', *Minutes of Proceedings of the Institution of Civil Engineers*, 24 (1865), pp. 280-314
Beaver, Patrick, *The Crystal Palace, 1851-1936: a Portrait of Victorian Enterprise* (Chichester, 1989)
Berman, Marshall, *All that is Solid Melts into Air: the Experience of Modernity* (London, 1983)
Bernheimer, Charles, 'Of Whores and Sewers: Parent-Duchâtelet, Engineer of Abjection', *Raritan: A Quarterly Review*, 6: 3 (1987), pp. 72-90
Biddle, Gordon, *Britain's Historic Railway Buildings: an Oxford Gazetteer of Structures and Sites* (Oxford, 2003)
Biswas, Asit K., *A History of Hydrology* (Amsterdam, 1970)
Blau, Eve, *Ruskinian Gothic: the Architecture of Deane and Woodward, 1845-61* (Princeton, New Jersey, 1982)
Bobrick, Benson, *Labyrinths of Iron: a History of the World's Subways* (New York, 1981)
Booker, Peter J., *A History of Engineering Drawing* (London, 1979)
Booth, George R., *The London Sewerage Question. Some Serious Objections and Suggestions Upon the Defective Plan of Sewerage Proposed by the Metropolitan Board of Works, Together with a Method for Remedying the Evil* (London, *c.*1853)
Boulton, J.J. (editor), *Edmund Burke: a Philosophical Enquiry into the Origins of Our Ideas of the Sublime and Beautiful* (London, 1958)
Bourne, Henry R.F., *English Newspapers: Chapters in the History of Journalism, vol. II* (Chippenham, 1998)
Briggs, Asa, *Iron Bridge to Crystal Palace: Impact and Images of the Industrial Revolution* (London, 1979)
Bright, Michael, *Cities Built to Music: Aesthetic Theories of the Victorian Gothic Revival* (Columbus, Ohio, 1984)
Brindle, Steven, *Paddington Station: its History and Architecture* (London, 2004)
Brooks, Chris, *Signs for the Times: Symbolic Realism in the mid-Victorian World* (London and Boston, 1984)
Brooks, Michael W., *John Ruskin and Victorian Architecture* (New Brunswick and London, 1987)
Brown, John K., 'Design Plans, Working Drawings, National Styles: Engineering Practice in Great Britain and the United States, 1775-1945', *Technology and Culture*, 41: 2 (2000), pp. 195-238
'When Machines Became Grey and Drawings Black and White: William Sellers and the Rationalization of Mechanical Engineering', *IA, The Journal of the Society for Industrial Archaeology*, 25 (1999), pp. 29-54
Brundage, Anthony, *England's 'Prussian Minister': Edwin Chadwick and the Politics of Government Growth 1832-54* (London, 1988)
Buchanan, Robert A., 'The Rise of Scientific Engineering in Britain', *British Journal for the History of Science*, 18: 2 (1985), pp. 218-33

The Engineers: a History of the Engineering Profession in Britain, 1750-1914 (London, 1989)

Bullen, J.B., *Byzantium Rediscovered: the Byzantine Revival in Europe and America* (London and New York, 2003)

Burdett Hederstedt, Henry, 'An Account of the Drainage of Paris', *Minutes of Proceedings of the Institute of Civil Engineers*, 24 (1865), pp. 257-79

Bynum, William F., and Roy Porter (editors), *Living and Dying in London* (London, 1991)

Carey, Peter, *Jack Maggs* (London, 1997)

Carls, Kenneth, and James Schmiechen, *The British Market Hall: a Social and Architectural History* (New Haven and London, 1999)

Certeau, Michael De, *The Practice of Everyday Life*, trans. Steven Rendall (London and Berkeley, California, 1984)

Cheney, Mary, *The Horton General Hospital, a Record of 100 Years of Service, 1872-1972* (Banbury, 1972)

Cherry, Bridget, and Nikolaus Pevsner, *The Buildings of England: London 4: North* (London, 1998)

The Buildings of England: London 2: South (London, 2001)

Cherry, Bridget, Ian Nairn, and Nikolaus Pevsner, *The Buildings of England: Surrey* (New Haven and London, 2002)

Cherry, Bridget, Charles O'Brien, and Nikolaus Pevsner, *The Buildings of England: London 5: East* (New Haven and London, 2005)

Clark, Clare, *The Great Stink* (London, 2005)

Clarke, George Rochfort, *The Reform of Sewers: Where Shall we Bathe? What Shall we Drink? Or, Manure Wasted and Land Starved* (London, 1860)

Clayton, Antony, *Subterranean City: Beneath the Streets of London* (London, 2000)

Clifton, Gloria, *Professionalism, Patronage and Public Service in Victorian London* (London, 1992)

Close, Charles, *The Early Years of the Ordnance Survey* (Newton Abbot, 1969)

Clutterbuck, R.H., 'Some Account of the Abbey of West Ham, Otherwise Stratford Langthorne', *Transactions of the Essex Archaeological Society*, 2 (1863), pp. 112-21

Connolly, Thomas W.J., *The History of the Corps of Royal Sappers and Miners*, 2 vols (London, 1855)

Cooney, E.W., 'The Building Industry', in Roy Church (editor), *The Dynamics of Victorian Business: Problems and Perspectives to the 1870s* (London, 1970), pp. 142-60

Corbin, Alain, *The Foul and the Fragrant: Odor and the French Social Imagination* (Cambridge, 1986)

Cottingham, Lewis N., *The Smith's, Founder's, and Ornamental Metal Worker's Director* (London, 1824)

Crook, J. Mordaunt, *The Dilemma of Style: Architectural Ideas from the Picturesque to the Post-Modern* (London, 1989)

Curl, James Stevens, *Victorian Architecture: its Practical Aspects* (Newton Abbot, 1973)

Victorian Architecture: Diversity and Invention (Reading, 2007)

Curtis, Gerard, 'Ford Madox Brown's *Work*: an Iconographic Analysis', *Art Bulletin*, 74: 4 (1992), pp. 623-36

Darby, Michael, and David Van Zanten, 'Owen Jones's Iron Buildings of the 1850's', *Architectura*, 4 (1974), pp. 53-75

Darlington, Ida, 'Edwin Chadwick and the First Large-scale Ordnance Survey of London', *Transactions of London and Middlesex Archaeological Society*, 22 (1969), pp. 58-63

Darlington, Ida, and James Howgego, *Printed Maps of London, c.1553-1850* (London, 1964)

Dean, Ptolemy, 'Terminal Decline?', *Country Life*, 194: 36 (2000), pp. 190-1

De Maré, *The Victorian Woodblock Illustrators* (London, 1980)

Denvir, Bernard, *The Early Nineteenth Century: Art, Design and Society 1789-1852* (London and New York, 1984)

The Late Victorians: Art, Design and Society 1852-1910 (London and New York, 1986)

Dixon, Roger, and Stefan Muthesius, *Victorian Architecture* (London, 1985)

Dobraszczyk, Paul, 'Sewers, Wood Engraving and the Sublime: Picturing London's Main Drainage System in the *Illustrated London News*, 1859-62', *Victorian Periodicals Review*, 38: 4 (2006), pp. 349-78

'Historicizing Iron: Charles Driver and the Abbey Mills Pumping Station (1865-68)', *Architectural History*, 49 (2006), pp. 223-56

'"Monster Sewers": Experiencing London's Main Drainage System', in Niall Scott (editor), *Monsters and the Monstrous: Myths of Enduring Evil* (Amsterdam and New York, 2007), pp. 9-32

'Architecture, Ornament and Excrement: the Crossness and Abbey Mills Pumping Stations', *Journal of Architecture*, 12: 4 (2007), pp. 353-65

'Mapping Sewer Space in Mid-Victorian London', in Ben Campkin and Rosie Cox (editors), *Dirt: New Geographies of Dirt and Purity* (London, 2007), pp. 123-37

'Image and Audience: Contractual Representation and London's Main Drainage System', *Technology and Culture*, 49: 3 (2008), pp. 568-98

Drabble, Margaret, *The Middle Ground* (London, 1980)

Dresser, Christopher, *The Art of Decorative Design* (London, 1862)

Driver, Charles Henry, 'Engineering, its Effects upon Art', *Transactions of the Civil and Mechanical Engineers' Society* (1874), pp. 3-12

'On Iron as a Constructive Material', *RIBA Transactions First Series*, 25 (1875), pp. 165-83

'Engineering and Art', *Transactions of the Civil and Mechanical Engineers' Society* (1879), pp. 2-9

'Presidential Address', *Minutes of Proceedings of the Civil and Mechanical Engineers' Society*, 522 (19 December 1879), pp. 6-7

Eastlake, Charles Locke, *A History of the Gothic Revival* (Leicester, 1978)

Eastleigh, R., 'Where There's Muck There's ... a Masterpiece', *Architect & Surveyor*, 62: 6 (1988), pp. 18-20

Engen, Rodney K., *Dictionary of Victorian Wood Engravers* (Cambridge, 1985)

Feaver, William, *The Art of John Martin* (Oxford, 1975)

Felstead, Alison, Jonathan Franklin, and Leslie Pinfield, *Directory of British Architects, 1834-1914* (London, 1993)

Ferguson, Eugene, *Engineering and the Mind's Eye* (Cambridge, Massachusetts, 1992)

Fergusson, James, *The Illustrated Handbook of Architecture: being a Concise and Popular Account of the Different Styles of Architecture Prevailing in all Ages and Cultures* (London, 1855)

History of the Modern Styles of Architecture (London, 1862)

Finer, Samuel E., *The Life and Times of Sir Edwin Chadwick* (London, 1952)

Flinn, Michael W. (editor), *Report on the Sanitary Condition of the Labouring Population of Great Britain, Edwin Chadwick, 1842* (Edinburgh, 1965)

Flather, Henry, *London Metropolitan Railway from Paddington to Finsbury Circus. Photographical Views, to Illustrate Works in Progress, July 1862* (London, 1862)

Fox, Celina, 'Wood Engravers and the City', in Ira B. Nadel and F. S. Schwarzbach (editors), *Victorian Artists and the City: a Collection of Critical Essays* (Oxford, 1980), pp. 1-11

Francis, A.J., *The Cement Industry, 1796-1914: a History* (Newton Abbot, 1977)

Fraser, Derek, *Power and Authority in the Victorian City* (Oxford, 1979)

Garrigan, Kristene O., *Ruskin on Architecture: His Thought and Influence* (Madison, Wisconsin and London, 1973)

Gay, John, and Gavin Stamp, *Cast Iron: Architecture and Ornament, Function and Fantasy* (London, 1985)

Gayle, Margot, and Carol Gayle, *Cast-iron Architecture in America: the Significance of James Bogardus* (London and New York, 1998)

Giedion, Sigfried, *Building in France, Building in Iron, Building in Ferro-concrete* (Santa Monica, California, 1928)

Space, Time and Architecture: the Growth of a New Tradition (Cambridge, Massachusetts, 1952)

Gilbert, Patricia K., *Mapping the Victorian Social Body* (Albany, New York, 2004)

Glick, Thomas F., 'Science, Technology, and the Urban Environment: the Great Stink of 1858', in Lester J. Bilsky (editor), *Historical Ecology: Essays on Environment and Social Change* (Port Washington, 1980), pp. 122-39, 187-8

Gloag, John, and Derek Bridgewater, *A History of Cast Iron in Architecture* (London, 1948)

Goddard, Nicholas, 'Nineteenth-century Recycling: the Victorians and the Agricultural Utilisation of Sewage', *History Today*, 31 (1981), pp. 32-6

Goodman, David C., and Colin Chant, *European Cities and Technology: Industrial to Post-industrial City* (London, 1999)

Gowers, Emily, 'The Anatomy of Rome from Capitol to Cloaca', *Journal of Roman Studies*, 85 (1995), pp. 23-32

Graham, Malcolm, and Laurence Waters, *Banbury: Past and Present* (Stroud, 1999)

Graves, Algernon, *The Royal Academy of Arts: a Complete Dictionary of Contributors and Their Work from its Foundation in 1769 to 1904*, 8 vols (London, 1905)

Griffiths, Dennis (editor), *The Encyclopedia of the British Press* (Basingstoke, 1992)
Hadfield, Charles, *British Canals: an Illustrated History* (Newton Abbott, 1984)
Hall, Peter, *Cities of Tomorrow* (Oxford, 2000)
Halliday, Stephen, *The Great Stink of London: Sir Joseph Bazalgette and the Cleansing of the Victorian Metropolis* (Stroud, 1999)
Hamilton, Ellis, *British Railway History* (London, 1954)
Hamlin, Christopher, 'Providence and Putrefaction:Victorian Sanitarians and the Natural Theology of Health and Disease', in Patrick Brantlinger (editor), *Energy & Entropy: Science and Culture in Victorian Britain* (Bloomington, Indiana, 1989), pp. 92-123
'Edwin Chadwick and the Engineers, 1842-1854: Systems and Anti-systems in the Pipe-and-brick Sewers War', *Technology and Culture*, 33 (1992), pp. 680-709.
Public Health and Social Justice in the Age of Chadwick: Britain 1850-1854 (Cambridge, 1998)
Hardy, Anne, 'Parish Pump to Private Pipes: London's Water Supply in the Nineteenth Century', in William F. Bynum and Roy Porter (editors), *Living and Dying in London* (London, 1991), pp. 76-93
Harley, John B., *The New Nature of Maps* (Baltimore and London, 2001)
'Maps, Knowledge and Power', in Stephen Daniels and Denis Cosgrove (editors), *The Iconography of Landscape: Chapters on the Symbolic Representation, Design and Use of Past Environments* (Cambridge, 1988), pp. 277-312
Harper, Roger H., *Victorian Architectural Competitions: an Index to British and Irish Architectural Competitions in The Builder 1843-1900* (London, 1983)
Harris, Thomas, *Victorian Architecture: a Few Words to Show that a National Architecture Adapted to the Wants of the Nineteenth Century is Attainable* (London, 1860)
Harrison, Michael, *London Beneath the Pavement* (London, 1961)
Hartwick, Georg L., *The Subterranean World* (London, 1871)
Helps, Arthur, *Life and Labours of Mr Brassey* (London, 1969, original 1872)
Herford, C.H., Percy and Evelyn Simpson (editors), *Ben Johnson* (11 vols, Oxford, 1954-70), vol. 8
Hindle, Brooke, *Emulation and Invention* (New York, 1981)
Hipple, Walter J., *The Beautiful, the Sublime and the Picturesque in Eighteenth-century British Aesthetic Theory* (Carbondale, Illinois, 1957)
Hitchcock, Henry-Russell, *Early Victorian Architecture in Britain*, 2 vols (New Haven, Connecticut, 1954)
Hollingshead, John, *Underground London* (London, 1862)
Hope, Alexander J.B. Beresford, *The Common Sense of Art: a Lecture Delivered in Behalf of the Architectural Museum, 8 December 1858* (London, 1858)
Horan, Julie L., *The Porcelain God: a Social History of the Toilet* (London, 1996)
Horrocks, Claire, 'The Personification of "Father Thames": Reconsidering the Role of the Victorian Periodical Press in the "Verbal and Visual Campaign" for Public Health Reform', *Victorian Periodicals Review*, 36: 1 (2003), pp. 2-19

Houfe, Simon, *The Dictionary of British Book Illustrators and Caricaturists 1800-1914, with Introductory Chapters on the Rise and Progress of the Art* (Woodbridge, 1978)
Hugo, Victor, *Les Misérables* (London, 1976; original 1862)
Humber, William, (editor), *A Record of the Progress of Modern Engineering 1863* (London, 1863)
Hyde, Ralph, *Printed Maps of Victorian London, 1851-1900* (Folkestone, 1975)
Jackson, Mason, *The Pictorial Press, its Origins and Progress* (London, 1885)
Jennings, Mary-Lou, and Charles Madge (editors), *Pandemonium 1660-1886: the Coming of the Machine as Seen by Contemporary Observers* (New York, 1985)
Jephson, Henry, *The Sanitary Evolution of London* (London, 1907)
Johnson, E.D.H., 'The Making of Ford Madox Brown's *Work*', in Ira B. Nadel and F. S. Schwarzbach (editors), *Victorian Artists and the City: a Collection of Critical Essays* (Oxford, 1980), pp. 142-51
Jones, Edgar, *Industrial Architecture in Britain, 1750-1939* (London, 1985)
Jones, Owen, *The Grammar of Ornament* (London, 1856)
Lectures on the Decorative Arts (London, 1862)
Jordan, David P., *Transforming Paris: The Life and Labor of Baron Haussmann* (New York, 1995)
Kearns, G., 'Private Property and Public Health Reform in England, 1830-1870', *Social Science and Medicine*, 26 (1988), pp. 187-99
Klingender, Francis D., *Art and the Industrial Revolution* (London, 1968)
Kohlmaier, Georg, and Barna von Sartory, *Houses of Glass: a Nineteenth-century Building Type* (Cambridge, Massachusetts, 1991)
Lefebvre, Henri, *The Production of Space*, trans. David Nicholson-Smith (Oxford, 1991)
Lefèvre, Wolfgang, (editor), *Picturing Machines 1400-1700* (Cambridge, Massachusetts, 2004)
Lesser, Wendy, *The Life Below Ground: a Study of the Subterranean in Literature* (Boston, 1987)
Lewis, R.A., *Edwin Chadwick and the Public Health Movement, 1832-1854* (London, New York and Toronto, 1952)
Liebig, Justus von, *Chemistry in its Applications to Agriculture and Physiology* (London, 1842)
Lister, Raymond, *Decorative Cast Ironwork in Great Britain* (London, 1960)
Lloyd, David, and Donald Insall, *Railway Station Architecture* (Newton Abbot, 1967)
Lubar, Steven, 'Representation and Power', *Technology and Culture*, 36: 2 (1995), pp. 54-82
Luckin, William, *Pollution and Control: a Social History of the Thames in the Nineteenth Century* (Bristol, 1986)
Macfarlane, Walter, *Examples Book of Macfarlane's Castings* (Glasgow, 1874)
McGee, David, 'From Craftsmanship to Draftsmanship: Naval Architecture and the Three Traditions of Early Modern Design', *Technology and Culture*, 40 (1999), pp. 209-36

McMahon, Cliff, *Reframing the Theory of the Sublime: Pillars and Modes* (Lewiston and Lampeter, 2004)

Macready, Sarah, and Frederck H. Thompson, and Patrick Nuttgens (editors), *Influences in Victorian Art and Architecture* (London, 1985)

Maidment, Brian E., *Reading Popular Prints 1790-1870* (Manchester, 1996)

Markus, Thomas A., *Buildings and Power: Freedom and Control in the Origin of Modern Building Types* (London and New York, 1993)

Martin, G.H., and David Francis, 'The Camera's Eye', in H. J. Dyos and Michael Wolff (editors), *The Victorian City: Images and Realities*, 2 vols (London and Boston, 1973), vol. 1, pp. 227-46

Mayhew, Henry, *London Labour and the London Poor*, 4 vols (London, 1862)

Meeks, Carroll L.V., *The Railroad Station: an Architectural History* (New Haven, Connecticut, 1956)

Middlemas, R. Keith, *The Master Builders: Thomas Brassey; Sir John Aird; Lord Cowdray; Sir John Norton-Griffiths* (London, 1963)

Muthesius, Stefan, 'The "Iron Problem" in the 1850s', *Architectural History*, 13 (1970), pp. 58-63

The High Victorian Movement in Architecture 1850-1870 (London and Boston, 1972)

Nairn, Ian, and Nikolaus Pevsner, *The Buildings of England: Surrey* (New Haven and London, 2002)

Nead, Lynda, *Victorian Babylon - People, Streets and Images of Nineteenth-century London* (New Haven and London, 2000)

Newman, John, *The Buildings of England: West Kent and the Weald* (London, 2000)

Nye, David E., *American Technological Sublime* (Cambridge, Massachusetts, 1994)

Oettermann, Stephen, *The Panorama: History of a Mass Medium* (New York, 1997)

Osborne, Thomas, 'Security and Vitality: Drains, Liberalism and Power in the Nineteenth Century', in Andrew Barry, Thomas Osbourne and Nicholas Rose (editors), *Foucault and Political Reason: Liberalism, Neo-Liberalism and Rationalities of Government* (London, 1996)

Owen, David, *The Government of Victorian London, 1855-1889: the Metropolitan Board of Works, the Vestries, and the City Corporation* (Cambridge, Massachusetts, 1982)

Owen, Tim, and Elaine Pilbeam, *Ordnance Survey: Map Makers to Britain since 1791* (Southampton, 1992)

Parliamentary Debates, 3rd series, vol. 97, 1848

Parliamentary Papers (1834, 1844, 1847-8, 1852, 1852-3, 1854-5, 1857, 1888)

Pevsner, Nikolaus, and Bridget Cherry, *The Buildings of England: Northamptonshire* (Harmondsworth, 1973)

Pevsner, Nikolaus, and Enid Radcliffe, *The Buildings of England: Essex* (Harmondsworth, 1965)

Pevsner, Nikolaus, and Jennifer Sherwood, *The Buildings of England: Oxfordshire* (Harmondsworth, 1979)

Phillips, John F.C., *Shepherd's London* (London, 1976)

Piggott, Jan R., *Palace of the People: the Crystal Palace at Sydenham, 1854-1936* (London, 2004)

Pike, David L., 'Modernist Space and the Transformation of Underground London', in Pamela K. Gilbert (editor), *Imagined Londons* (New York, 2002), pp. 101-19

Subterranean Cities: the World Beneath Paris and London, 1800-1945 (Ithaca and London, 2005)

Pinkney, David H., *Napoleon III and the Rebuilding of Paris* (Princeton, New Jersey, 1958)

Pollard, Sidney, *The Genesis of Modern Management* (London, 1965)

Porter, Dorothy (editor), *The History of Public Health and the Modern State* (Amsterdam, 1994)

Porter, Dale H., *The Thames Embankment: Environment, Technology, and Society in Victorian London* (London, 1998)

Potts, Alex, 'Picturing the Metropolis: Images of London in the Nineteenth Century', *History Workshop*, 26 (1988), pp. 28-56

Pudney, John, *The Smallest Room* (London, 1954)

Rawlinson, Robert, *Designs for Factory, Furnace and Other Tall Chimney Shafts* (London, 1858)

Reid, Donald, *Paris Sewers and Sewermen: Representations and Realities* (Cambridge, Massachusetts, 1991)

Richardson, Ruth, and Robert Thorne, *The Builder Illustrations Index, 1843-1883* (London, 1994)

Robertson, Edward G., and Joan Robertson, *Cast Iron Decoration: a World Survey* (London, 1977)

Robinson, William, *Glastonbury Abbey, Somersetshire* (Wells, 1843)

Rodgers, Jasper W., *Fact and Fallacies of the Sewerage System of London and other Large Towns* (London, 1858)

Rolt, L.T.C., *Victorian Engineering* (London, 1970)

Ruskin, John, *The Seven Lamps of Architecture*, 2nd edition (Orpington, 1849)

Examples of the Architecture of Venice: Selected and Drawn to Measurement from the Edifices (London, 1851)

The Stones of Venice, 3 vols (London, 1851-53)

The Two Paths (London, 1904)

Saint, Andrew, *Architect and Engineer: a Study in Sibling Rivalry* (New Haven and London, 2007)

Sankey, H. Riall, and Ian Mumford, *The Maps of the Ordnance Survey: a Mid-Victorian View* (London, 1995)

Saunders, Ann, *The Art and Architecture of London: an Illustrated Guide* (Oxford, 1988)

Schivelbusch, Wolfgang, *The Railway Journey: the Industrialization of Time and Space in the Nineteenth Century* (Los Angeles, 1977)

Scott, George Gilbert, *Remarks on Secular and Domestic Architecture, Present & Future* (London, 1857)

Sheail, John, 'Town Wastes, Agricultural Sustainability and Victorian Sewage', *Urban History*, 23 (1996), pp. 189-210

Simmons, Jack, and Robert Thorne, *St Pancras Station* (London, 2003)
Sinnema, Peter W., *Dynamics of the Printed Page: Representing the Nation in the Illustrated London News* (Aldershot, 1998)
Skelton, R. A., 'The Origins of the Ordnance Survey in Great Britain', *Geographical Journal*, 128 (1962), pp. 415-30
Smith, Denis, 'Pumping Stations', *Architectural Review*, 151 (1972), pp. 324-8
'Sir Joseph Bazalgette (1819-1891): Engineer to the Metropolitan Board of Works', *Transactions of the Newcomen Society*, 58 (1986), pp. 89-111
Sir Joseph Bazalgette: Civil Engineering in the Victorian City (London, 1991)
Smith, J., 'Crossness', *Architectural Review*, 146 (1969), pp. 411-14
Smith, Raymond, and Nicholas Young, 'Sewers Past and Present', *History Today*, 43 (1993), pp. 8-10
Smith, Stephen, *Underground London: Beneath the Streets of the City* (London, 2004)
Stamp, Gavin, and Colin Amery, *Victorian Buildings in London, 1837-1887: an Illustrated Guide* (London, 1980)
Stevens, Edward W., *The Grammar of the Machine: Technical Literacy and Early Industrial Expansion in the United States* (New Haven and London, 1995)
Stevens, Frank L., *Under London: a Chronicle of London's Underground Life-lines and Relics* (London, 1939
Street, George Edmund, *Brick & Marble in the Middle Ages: Notes of a Tour in the North of Italy* (London, 1855)
Summerson, John, *The London Building World of the Eighteen-sixties* (London, 1973)
The Architecture of Victorian London (Charlottesville, Virginia, 1976)
Sunderland, David, '"A Monument to Defective Administration"? The London Commissions of Sewers in the Early Nineteenth Century', *Urban History*, 26: 3 (1999), pp. 349-72
Taylor, Nicholas, 'The Awful Sublimity of the Victorian city: its Aesthetic and Architectural Origins', in H. J. Dyos and Michael Wolff (editors), *The Victorian City: Images and Realities*, 2 vols (London and Boston, 1973), vol. 2, pp. 431-47
Thompson, Francis M.L., *Chartered Surveyors: the Growth of a Profession* (London, 1968)
Thompson, Paul, *William Butterfield* (London, 1971)
Trench, Richard, and Ellis Hillman, *London Under London: a Subterranean Guide* (London, 1984)
Turner, Thomas H., *Some Account of Domestic Architecture in England from the Conquest to the End of the Thirteenth Century* (Oxford, 1851)
Van Zanten, David, *The Architectural Polychromy of the 1830s* (London, 1977)
Wadsworth, Cliff, *George Furness: Willesden's Greatest Resident* (London, 1996)
Walford, Edward, *Greater London: a Narrative of its History, its People, and its Places*, 2 vols (London, 1897)
Walker, Charles, *Thomas Brassey: Railway Builder* (London, 1969)
Walker, Derek., *The Great Engineers: the Art of British Engineers 1837-1987* (New York, 1987)

White, William, 'Ironwork: its Legitimate Uses and Proper Treatment', *RIBA Transactions,* 1st series, 16 (1865), pp. 15–30

Wiggins, John, *The Polluted Thames: the Most Speedy, Effectual, and Economical Mode of Cleansing its Waters, and Getting Rid of the Sewage of London* (London, 1858)

Williams, Rosalind, *Notes on the Underground: an Essay on Technology, Society and Imagination* (Cambridge, Massachusetts, 1990)

Williamson, John, *Glastonbury Abbey: its History and Ruins* (Wells, 1858)

Wilson, Aubrey, *London's Industrial Heritage* (Newton Abbot, 1967)

Wilton, Andrew, *Turner and the Sublime* (London, 1980)

Wohl, Anthony S., *Endangered Lives: Public Health in Victorian Britain* (London, 1973)

Wright, Lawrence, *Clean & Decent: the Fascinating History of the Bathroom and the Water-closet* (London, 1960)

Wyatt, Matthew Digby, 'Iron-work and the Principles of its Treatment', *Journal of Design and Manufactures*, 4: 19 (1850), pp. 10–14; and 4: 21 (1850), pp. 74–8

Metal-work and its Artistic Design (London, 1852)

Wyatt, Matthew Digby, and John Burley Waring, *The Byzantine and Romanesque Court in the Crystal Palace* (London, 1854)

Index

Page numbers in bold indicate illustrations. Information in notes is indicated in the form 199 n.23, i.e. note 23 on page 199.
Index compiled by Margaret Binns.